Physiology of
Animal Growth & Bioenergetics

Physiology of Animal Growth & Bioenergetics

U. K. MISHRA
M.V.Sc. & A.H., Ph.D., Post Doc. (Canada)
Professor and Head
Department of Veterinary Physiology
COLLEGE OF VETERINARY SCIENCE & ANIMAL HUSBANDRY,
ANJORA, DURG (CHHATTISGARH), INDIA-491 001

2008

SSPH

SATISH SERIAL PUBLISHING HOUSE

403, Express Tower, Commercial Complex, Azadpur, Delhi-110033 (INDIA)
Phone : 011-27672469, 27672852, Fax : 91-11-27672046
E-mail : info@satishserial.com, hkjain1975@yahoo.com
Visit us at : www.satishserial.com

Published by :

SATISH SERIAL PUBLISHING HOUSE
403, Express Tower, Commercial Complex, Azadpur,
Delhi-110033 (India)
Phone : 011-27672852, Fax : 91-11-27672046
E-mail : info@satishserial.com, hkjain1975@yahoo.com
Website : www.satishserial.com

First Published 2008

ISBN : 81-89304-46-1

Printed at

SALASAR IMAGING SYSTEMS, DELHI-35

PREFACE

The objective of this text book of physiology of animal growth and bioenergetics is to present an account on the basic principles of physiology pertains to the animal growth and bioenergetics. Though the books on growth physiology are available but on bioenergetics the only genuine book is brody's book entitled bioenergetics in growth which was published way back in 1945, since then this book has not been revised.

As a teacher of Veterinary Physiology for a long time I have noticed that the students of graduate and post graduate level do not find the concise information on growth and bioenergetics in one book and that to in a scattered manner. Due to this very fact the idea of writing the present book came.

In this book I don't pretend to claim any credit for original contribution. It is the concise collection of the important aspects from different comprehensive books written by the pioneer authors and some current publications.

I am indebted to all the editors, authors and publishers whose original findings have been incorporated in this book.

I would like to acknowledge the valuable help rendered by Dr. M.K.Gendle and Dr. Meenu Dubey, Ph.D. Students in the Deptt. of Animal Nutrition, College of Veterinary Sci. & AH., Anjora, Durg.

I am particularly thankful to M/s Satish Serial Publishing House, Delhi for Publishing of this book. Any suggestion from the student and teacher for its improvement will be highly appreciated for future revision of the text.

(U.K. Mishra)

LISTS OF FIGURES

LIST OF TABLE

CONTENTS

Chapter 1

Growth

The life of the organism begins from single cell known as zygote, which is less than 1/700th inch in diameter and so small in size that it can not be weighed accurately.

Through the process of cell division and differentiation, this cell develops into a mature individual weighing about 150 to 200 lbs and possessing about 250 trillions of cells (mature human beings).

Growth is biologic synthesis that is the interaction of exogenous materials (food) in the formation of new biochemical units.

Schloss (1911) defines growth as a "Correlated increase in the mass of the body in definite intervals of time in a way characteristic of the species." This definition implies that subject to individual variability there is a characteristic rate of growth for each species and a characteristic adult size and development.

Brody (1945) defines growth as "a relatively irreversible time change in the measured dimensions." This definition covers growth in the size as well as in weight and the concept of irreversible excludes flotation due to variation in food supply and to gestation and lactation.

Growth may be subdivided into

1. Cell multiplication *(Hyperplasia)*
2. Intussusception or increase in size of the cell *(Hypertrophy)*
3. *Accretion* or increase in amount of non-living structural matter.

The cleavage of the egg into daughter cells is growth by *cell multiplication*. Blood corpuscles, monocytes, hair follicle cells and ectoderm cells always grow by cell multiplication.

Nervous tissue and skeletal muscle tissue grow by *cell enlargement*, not by cell division.

The inclusion of non-protoplasmic substances such as fat, blood plasma, cartilage etc. is an increase by incorporation of material from the environment, involving neither cell multiplication nor cell enlargement.

Prenatal Growth

Prenatal growth results from a series of orderly differential process that transform a single celled zygote into an animal. The period of growth occurs between the time the ovum is fertilized and the young are born.

Pattern and Mechanism of Prenatal Growth

(A) Cleavage and Differentiation

After fertilization, the zygote divides and re-divides several times without increasing the volume of cytoplasm. The process of cellular division without an increase in volume is called *cleavage* and the cleaving cells are called blastomeres. The early cleavages are synchronous and occur at specific times after fertilization. During later cleavage, cell division is asynchronous and different parts of the embryo may cleave at different rates. When the dividing cells reached a certain number that is characteristics of the species, they are arranged in a one cell thick layer around a blastocoelic cavity the embryo then called a *blastocyst*.

Differentiation involves the progressive specialization of cells both structurally and functionally. During the first week of gestation, the blastocyst differentiates into two parts

1. *Trophoblast*: concern with the establishment of placenta.
2. *Inner cell mass* (embryonic disc) that develop into the embryo proper.

The blastocyst then becomes a *Gastrula*, which is successively two layered and three layered. *Ectoderm* (Primary outer) and *Endoderm* (inner) layer differentiate during the two layered stage. A third layer, the *Mesoderm*, then develops between this from the primitive streak (ectoderm). By various modifications, these three primary germ layers give rise to the organs of the embryo. Further developmental process converts the undifferentiated embryonic cells into the histological type peculiar to each organ. Cells can develop into more than one type of tissue before gastrualation. Thereafter, their fate is determined at the end of gastrulation. The environment of embryonic cells then determines the tissue into which they will differentiate. However, cells cannot be influenced or induced by surrounding tissue.

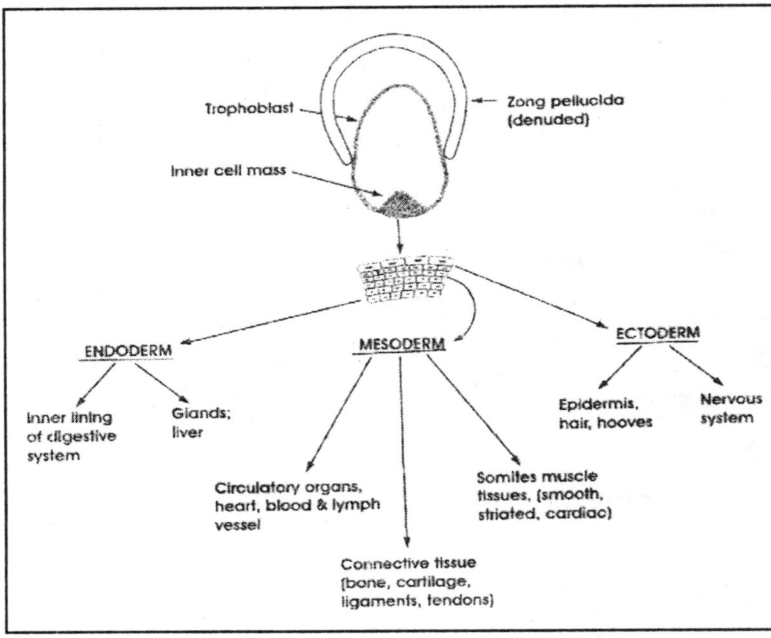

Fig. 1. Three primary germ layers and structures formed from them.

(B) Organogenesis (organ formation)

The organization of various dividing cells into special organs with a particular makeup is known as morphogenesis or organogenesis. Organogenesis refers to the establishment of organs rudiments through the interaction of germ layers. Organs are first blocked out in their general forms and then molded and shaped in detailed by process such as *Invagination* (in folding), *Exvagination* (out folding), and *Budding* or hollowing out. Cells from each of the three germ layer aggregate into primordial cell masses from which special organs will ultimately be formed.

The innermost layer (*Endoderm*) forms the inner lining of the gut, its glands and the bladder.

The outermost layer (*Ectoderm*) forms a mid dorsal ridge which gives rise to the brain, spinal cord, and other derivatives of the nervous system.

The *Mesodermal* layer gives rise to muscle, bones, connective tissue and vascular system. After the initiation of organogenesis, there is a period of rapid increase in organs size. This increase in dimension, which results from cells proliferation and increase in cell size, is the growth process.

Body organs are formed in an orderly sequence. The head portion of the embryo developed before the tail and cephalochordial (Head of spinal column) develops before other organs.

A series of somites (body segments) developed along each side of the spinal cord from the outer layer of the mesoderm. In cattle differentiation of somites starts on 19th day of pregnancy and number of somites are increased from 25 to 55 on 32nd day of pregnancy. Prenatal growth in form animals proceeds at varying rates among different species e.g. pigs, spend only 110 to 115 days in mother uterus whereas, foals spend 335 to 345 days. All farms mammals are born at a fairly comparable degree of maturity, in spite of difference in the pregnancy period in various species. Certain other animals are born still immature e.g. Opossum is born 12 days after conception. The young once are very immature at birth, since they did not have fetal membranes. This baby opossum are kept in the pouch where they suck the milk and develop also.

Baby rat is born 21 to 22 days after conception. At birth it is quite helpless, has no hairs, eyes are closed and unable to walk and run almost immediately.

Human infant is carried in uterus about 9 months and is helpless at birth even though its eyes are readily open and senses soon start functioning. Babies can't walk until they are nearly one year of age. The human female baby exceeds male in biological growth rate at early prenatal life. Thus full term male is more immature then females at birth. Male babies are most susceptible for death and thus, it appears that sex hormones are involved in the degree of maturity of human baby at birth.

The size of the young one at birth is determined by the genes from the sire but uterine environment of the mother has been found to be more important for the size of the fetus at birth e.g. *Small mare x large stallion=small size young*

Big mare x small stallion=large size young

After organogenesis the size of the organ develop due to fast multiplication of the cells (morphogenesis) and attain a definitive character. The number of the muscle fibers that individual possess are fixed at early fetal stage, which thereafter increase in size by hyperplasia. Skeletal size may be modified to certain extent by the environmental factors but genetic constitution plays an important role.

(C) Measurement of prenatal Growth

The size of the fetus is judged by total fetal length, crown- rump length, curved crown rump length or fetal weight. During prenatal development several growth curves can be employed for the fetus as a whole or for different organs. (Fig.2 and 3).

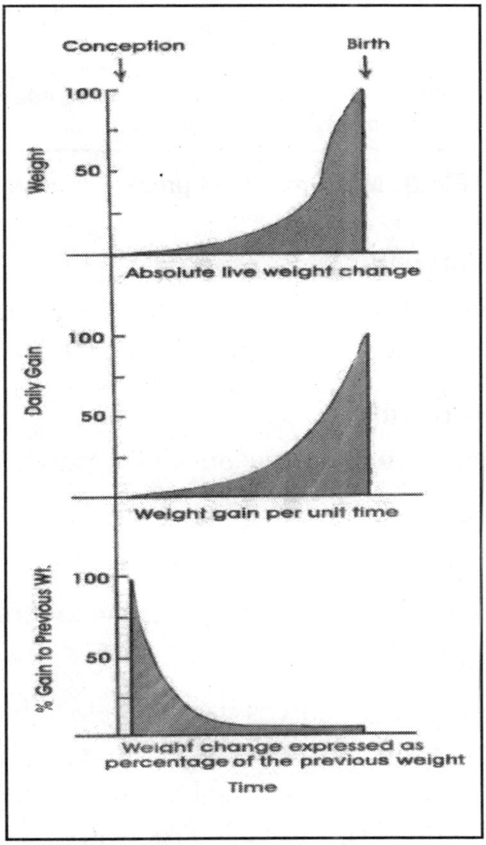

Fig. 2 Prenatal growth curve

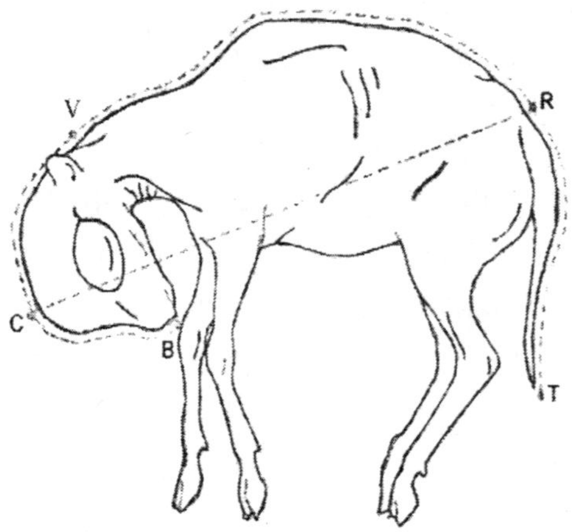

Total lenght BCVRT, Crown Rump length CVR, vertebral column length VR, Vertebral column and tail length VRT

Fig.3. Measurement of prenatal growth

i) Absolute growth

Absolute growth is the change in volume, crown rump length or weight of the fetus per unit time.

ii) Relative growth

Relative growth is the absolute growth per initial dimension.

iii) Specific growth

In the curve of specific growth, the logarithm of size is plotted against age.

Prenatal growth curves vary in details from species to species, but they resemble each other mathematically. Changes in the rate of prenatal growth during gestation indicates that fetal size is itself a regulator of growth i.e., the prenatal growth is in some measure, self regulating.

(D) Differential Growth

The birth weight of the fetus is approximately 60% of the total weight of the conceptus, the fetal membrane and placental fluid makeup the remaining 40%. Different parts of the conceptus grow at different rates during gestation. The fetal membrane grows rapidly during the early stage of gestation and their absolute weight reaches maximum during mid pregnancy. In the early stage of the embryonic development the cephalic region grows rapidly, so that the head of the fetus is consequently disproportionately large. During the later stage of the fetal life, when growth in the cephalic region is less rapid the proportion of the neonates are established. The growth gradients of the fetus follow a definite order for example at birth head, limbs and fore quarters are relatively, more developed then the muscles. Differential growth may be related to the functional needs, rapid growth of the mesodermal elements which forms the embryos, hearts is correlated with the necessity for circulating blood and nutrients to all parts of the growing fetus.

In the early stage of prenatal growth, muscle cells increase in number (Hyperplasia). During the later stage of prenatal and in post natal life the increase in the weight of the body muscles is due to an increase in the size of muscle cells (Hypertrophy). Skeletal growth of the fetus is constant, although some dimension increase more rapidly than others.

Fetal Hormone and Prenatal Growth

The endocrine glands viz. Pituitary, adrenal and thyroid glands are major glands of the fetus, which affect the fetal growth. Gonads to certain extent are also involved. Hypothalamus of the fetus is active and secretes various releasing hormones for the secretion of anterior pituitary hormones. One such important releasing hormone is GHRH (Growth hormone releasing hormone) and GHRIH (Growth hormone release inhibitory hormone) which in turn affect the STH (Somatotropic hormone) secretion from Anterior Pituitary. Hypothalamus also produces the CRH (Corticotrophin releasing hormone) and TSH (Thyroid stimulating hormone), which affect anterior pituitary to produce ACTH (Adrenocorticotropic hormone) and TSH.

Thyroid gland and gonads of the fetus are required for the morphogenesis. Thyroid deficiency prevents metamorphosis in frogs. In man and rat brain, development is retarded in thyroid deficiency.

Factors Affecting Prenatal Growth

1. Heredity

The ultimate size of the fetus is determined by its own genotype. However, maternal contribution to the fetal size is more than paternal contribution.

2. Size and age of dam

Faster prenatal growth is directly related to large maternal size. Maternal tissue competes with the fetal tissue for growth for longer period and consequently has more effect on the fetal size. Parity of mature dam in the beginning has positive effect on growth and subsequently negative.

3. Maternal Nutrition

The fetus is privileged from a nutritional point of view. Even with severe maternal undernourishment, it continues to grow and will achieve birth weight which is close to normal. Under certain circumstances, the growth of the fetus is proportional to the caloric intake of the mother. Providing good nourishment during third trimester is essential for better growth of fetus. However, good feeding in early pregnancy and poor feeding during late pregnancy causes poor growth of the fetus. Poor maternal nutrition during late gestation causes a reduction in the glycogen content of fetal muscles and liver. The fetal glycogen store serves as a source of energy immediately after birth. Thus, poor maternal nutrition may increase neonatal mortality. Certain fetal organs, which mature early, will suffer due to nutritional stress. Baby of diabetic lady will be healthy and sometime overweight due to passage of more glucose from mother to fetus through placenta.

4. Litter Size

In polytocous species, increased litter size reduces the rate of prenatal growth because of variation in the function of the placenta and duration of pregnancy. This reduction in the prenatal weight is due to limited availability of the nutrients. In monotocous species, multiple births usually retard prenatal growth.

5. Placental size

The placenta is often regarded as a fetal organ. Since all nutrients, which reach the fetus, are transported by the placenta, it is often assumed that prenatal growth is determined by the placenta and that a small placenta retards prenatal growth. The stunted growth of neonate piglets, often seen in large litters, is related to the small size of the placenta.

6. Ambient temperature

High ambient temperature during pregnancy affects fetal size in some species. Heat stress reduces fetal growth in ewes. Dwarfing of the lambs is due to prolonged exposure of ewes to high ambient temperature rather than reduced feed intake, which occurs during pregnancy. Animals, which are undersized at birth, are physiologically premature and are more subject to neonatal mortality owing to their poor thermoregulatory mechanism and their inability to withstand stress in a new environment.

—★★★—

Chapter 2

Postnatal Growth

Growth after birth usually begins slowly, then goes through a more rapid period and then slows down again or even may stop.

Neonatal Growth

The growth rate from birth to weaning may be greatly influenced by the amount of milk secreted by the mother and by the individual's health. Total growth during this period also depends upon the individual's inherent growth impulse. Growth of body parts after birth proceeds at different rates in most farm animals.

Immediately after birth, the bones generally grow rapidly than other tissues in proportion to the total body weight. This rapid skeletal growth decreases by weaning time. Maximum muscular growth occurs after skeletal growth and it continues to increase by weaning time.

The body weight increase after skeletal and muscular growth is due to the fat deposition. Adequate supply of nutrients hastens the growth and inadequate nutrition delays peak muscle growth and slows the rate of fat deposition.

Growth Curve

When body weight of an animal is plotted against the age the curve so obtained is known as growth curve which is sigmoid or roughly "S" shape.

The curve of growth may be divided into two principal segments,

a. Self accelerating phase

b. Self inhibiting phase.

Thus the shape of the growth curve determined by two opposing forces.

1. A growth accelerating force

2. A growth retarding force.

In the absence of inhibiting forces, the number of new individual produced per unit time is always proportional to the number of reproducing units, but there comes a time marked by the inflection in the growth curve when the increase in the population is not proportional to the number of reproducing individual in the population but to the available resources necessary for the growth. The resources may be in the form of food, space and freedom from the encroaching byproduct of growth for example lactic acid in the case of growth of lactic acids bacterium in the milk. It is the environment or the growth inhibiting forces in the environment that exerts the limiting influence on growth after the inflection in the curve.

The *Inflection* in the growth curve represents the position, at which the increase in the growth velocity ceases, and the decrease in the velocity has not yet begins, therefore the inflection represents the position at which gains are most rapid and more economical.

The point of Inflection Indicates

1. The time of maximum velocity of growth.
2. The age of puberty.
3. The lowest specific mortality.
4. A geometric referent for the determination of equivalence of age in different animals

The point of inflection occurs at about 14 years in man, 5 months in cattle and 2 months in sheep.

The shape of all growth curve, whether of individual or population, except man is sigmoid. In case of man the juvenile period is exceptionally long. In most of the higher animals puberty occurs after about 30 per cent of the mature weight has been achieved whereas in man it does not occur until about 60 or 70 per cent of the mature weight has been achieved the early phase is of rising slope and the later phase of declining slope. The junction between two phases occurs during puberty in animals, flowering in plants.

The lower asymptote is the starting level.

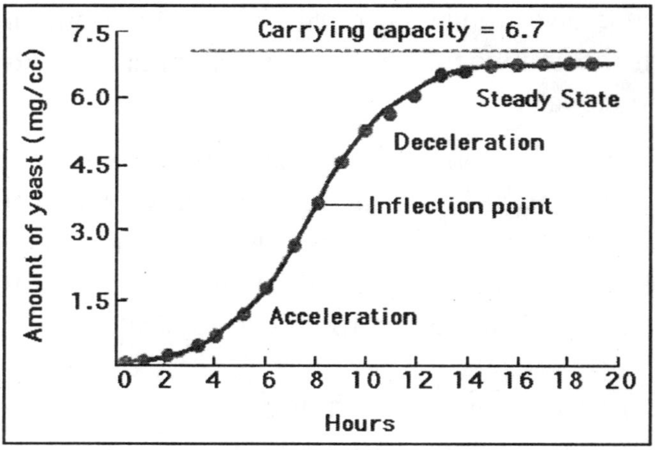

Fig. 4. A typical Growth Curve

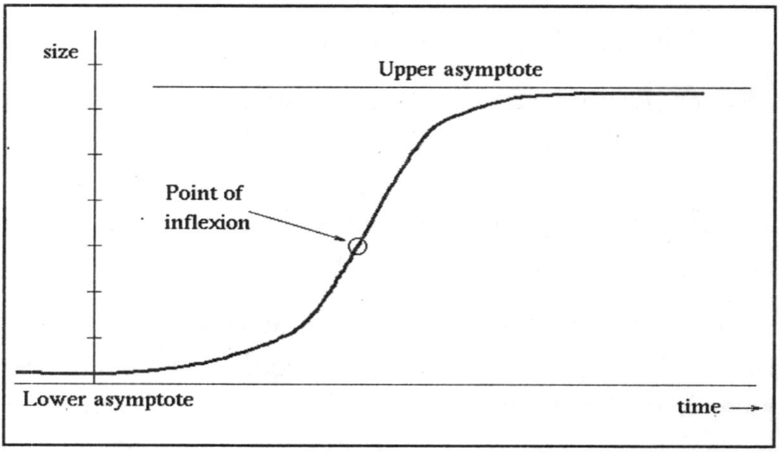

The upper asymptote is the mature level.

The point of inflexion is the point of maximum growth.

Measures of Growth

The commonest measure of growth is the increase in the live weight; however, various measures of size such as height and length are also used. A combination of live weight and measures of size is more informative than live weight alone, as it records the changes in the shape which occurs during normal growth and also demonstrates that an animal may continue to increase in size even though the body weight is kept constant.

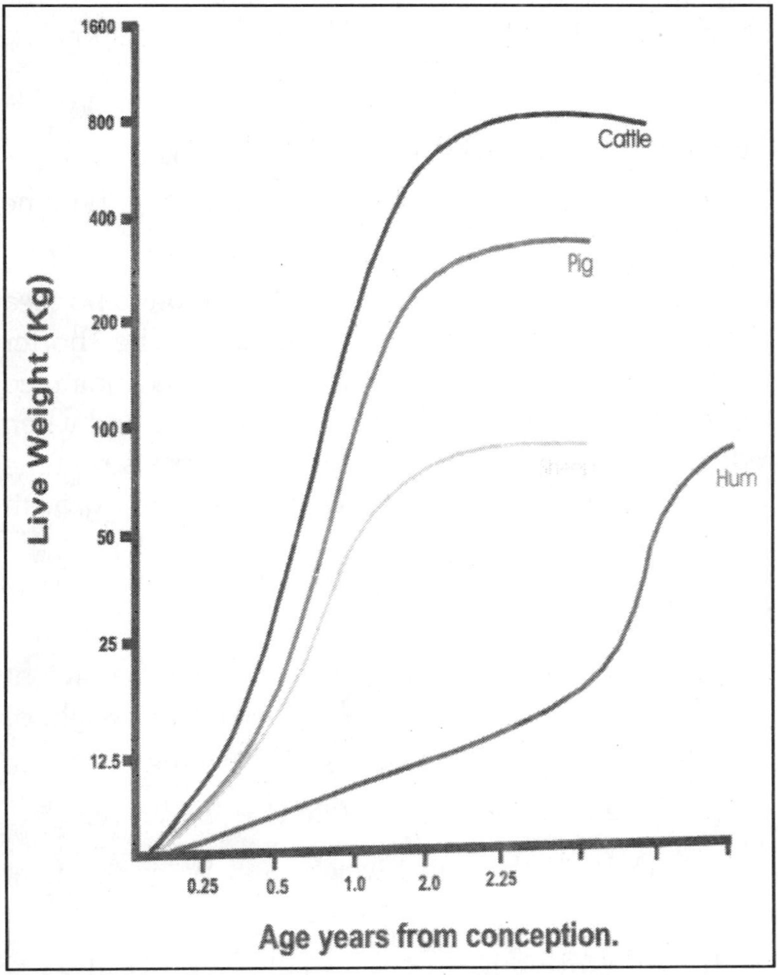

Fig. 5. Growth curve of major farm animals and human

Growth in weight is usually represented as

1. Absolute growth rate
2. Relative growth rate

Absolute Growth Rate

Absolute growth rate is the absolute gain in the given magnitude per unit time. The absolute gain in the observed weight difference W2-W1 for the corresponding time difference t_2-t_1 may be represented as

Average Absolute Growth Rate = $W_2-W_1/\ t_2-t_1$

Where W_2-W_1 is gain in the weight during the time interval t_2-t_1

If t_2-t_1 is long the average growth rate gives no idea of the growth rate at any particular time. The shorter the interval of time for which the average is computed the more nearly it approaches the true value and when reduced to an interval dt, so short that there is no time for the velocity of growth to change the true growth rate dw/dt is obtain.

Relative Growth Rate

Relative growth rate is represented by the weight gain during a given time interval divided by the weight of the organism W_1 at the beginning of the time.

Average Relative Growth Rate R= W_2-W_1/W_1

Where W_1 is the initial weight and W_2 is the final weight.

Relative growth rate is usually expressed as a percentage and is useful for comparing growth in different species of widely different body weight. If the gain in weight is very large in comparison with the initial weight, the relative growth rate gives no idea of the growth rate at the time when the final weight was recorded.

Instantaneous growth rate

The instantaneous growth rate is obtain by dividing the instantaneous weight gain dw/dt by the weight W which is the weight at which dw/dt is measured

Instataneous relative growth rate k = (dw/dt)/ (W)

Practically it is impossible to measure the instantaneous weight gain dw/dt. So the formula is written in the integrated form is

$$W = Ae^{kt}$$

Where e is the base of natural logarithm and A is the natural logarithm of W when t = 0

This formula is applicable to the self-accelerating phase of growth when the relative growth rate is a function of growth already made. During the self-inhibiting phase of the growth curve the relative growth rate is a function not of growth already made but of growth, yet to be made to reach maturity.

$$dw/dt = -k\,(A-W)$$

Where A is the mature weight, the integrated form of this equation is

$$W = A-B\,e^{-kt}$$

Where A is the mature weight and B is the constant.

Factors Affecting Live Weight Growth

1. Plane of Nutrition

The effect of plane of nutrition on live weight growth is important. Changes in the growth rate of meat animals due to changes in plane of nutrition directly affect the turn over and efficiency of conversion of food into meat. Varying plane of nutrition at different stages of development also affect the composition and therefore the value of the carcass. The ultimate effect on body weight of a period of retarded growth depends on three main points

1. The stage of growth at which retardation is applied.
2. Severity of retardation.
3. Length of time during which the retardation is imposed.

Nutrition of the female during last stage of pregnancy has a marked effect on milk yield largely through its influence on the growth of the udder (Wallace, 1948). Under- nutrition of the ewe during pregnancy leads to the production of under- sized and weak lambs even though the nutrition of the ewe earlier in pregnancy was on a high plane. Conversely, a high level of nutrition of the ewe in late pregnancy resulted in normal sized lambs even though the level of nutrition of the ewe in early pregnancy had been low.

2. Sex

The effect of sex on live-weight growth is two-fold. First, there is a direct effect of sex on growth resulting presumably from genetic differences between males and females, and secondly there is an indirect effect of sex due to the influence of the sex hormone. In most mammals and birds, males tend to grow faster and reach heavier weight than females. In horses, there is no appreciable difference between the sexes neither at birth nor for some time after birth. In sheep, the difference in growth rate between the sexes in sheep is apparent only when the nutritional regime on which the animals are reared is adequate for normal growth and when lambs were reared on a low plane of nutrition, there was virtually no difference between the growth rate of males and females. In poultry, there is no difference in weight between the sexes at birth but cockerels grow faster than hens and to a greater mature weight.

3. Hormones

Hormones do not act independently but as part of a complex pattern. Following hormones influences the growth of individuals.

a. Sex Hormones

The difference in growth rate between sexes is present not only post pubertally but also pre pubertally and even prenatally. Administration of oestrogens to normal females for prolong period inhibit growth of the long bone, which is because of ossification of the epiphyseal cartilage. Administration of Progesterone resulted in increased growth. With fattening lambs scientist found that stilboesterol, implanted sub cutaneously improved the rate of live weight gain and FCE. However, the dressing percentage and carcass quality were consistently reduced.

b. Thyroid

In young animals removal of the thyroid has been shown to retard growth in weight and the mature weight of thyroidectomized individual has been shown to be only half that of normal adults. Hypothyroidism is associated with low metabolic rate, reduced feed intake, low blood sugar and liver glycogen and low nitrogen retention. There is a tendency towards increased deposition of fat. The effect of hyperthyroidism is to increase metabolic rate and in severe hyperthyroidism, catabolism exceeds anabolism, so that the effect of hyperthyroidism is to produce a loss of body weight.

c. Adrenal Cortex

The cortex of the adrenal gland produces about 20 steroid hormones collectively known as Cortin. Lack of cortin produces profound hypoglycemia, loss of Na, Cl and water, retention of K, low blood pressure, weak heart action, reduced basal metabolism, disturbed temperature regulation and general muscular weakness. As this gland is so closely concerned with regulation of salt water balance and with carbohydrate and protein metabolism it has a profound effect on growth.

d. Anterior Pituitary

The pituitary gland is the most important of the growth regulators. It produces a series of trophic hormones which control the activity of other endocrine glands and hypophysectomy leads to atrophy of these i.e. thyroid, adrenal and Islets of Langerhans. Hypofunction of the Pituitary upset the whole metabolic process which in turn has an adverse affect on growth.

4. Vitamins

There has been little or no evidence of an effect on growth rate of a sub clinical vitamin deficiency (a state of vitamin deficiency which is not sufficient to produce a characteristics disease is sufficient to retard growth rate). In case of pigs and poultry, diets composed entirely of foods of vegetable origin were inadequate for producing maximum growth rate even though supplemented with all the known vitamins. When these diets were supplemented with animal protein food, growth rate were greatly improved and this was ascribed to the presence of an unidentified factor known as the animal protein factor.

5. Antibiotics

The antibiotics commonly used are Penicillin, Aureomycin, Streptomycin, Teramycin, Chloromycetin and Bacitracin. In all cases antibiotics improve the growth rate of pigs and poultry on an all vegetable diet, but the response of animals on diet containing animal protein supplement is less certain and appears to depend on some extent on the condition under which the animals are kept. The exact mode of action of antibiotics in stimulating growth has not yet been fully worked out but it seems that the effect is produced through their influence on the microflora of the gut. Growth stimulation is greatest in unthrifty animals kept in unhygienic surroundings suggest that the effect of antibiotics is to combat actively pathogenic organism which will not actually producing clinical signs of disease are produced toxins which have a growth depressant effect.

—★★★—

Chapter 3

Growth Regulators

Substances, which have a substantial and direct effect on growth, are classified as growth regulators. This includes:

I. Tissue specific growth factor

II. Certain hormone

Tissue Specific growth factor acts as cellular growth regulators by affecting the proliferation and development of specific types of cells. These include epidermal growth factor, nerve growth factor, erythropoietin, promine, retine and tissue specific chalones.

I. Cellular Growth Regulators

a. Epidermal Growth Factor (EGF)

Epidermal growth factor stimulates cell proliferation in vivo and in vitro. Injection of purified EGF into mice, chicks, rats and some other animals increases the dry weight, total nitrogen, DNA, RNA and some enzymes of the epidermis. Epidermal growth factor acts directly on the epidermis.

b. Nerve Growth Factor (NGF)

Potent source of NGF were found to be in mouse sub-maxillary gland and snake venom. It has little species specificity but is highly specific for the type of nerve tissue that it stimulates. Injection of NGF into chick embryos cause massive invasion of viscera and blood vessels by sympathetic nerve fibres.

c. Erythropoietin

Erythropoietin is a hormone circulating in the blood, which stimulates the formation of Red Blood Cells. The production of Erythropoietin increases in response to anaemia, hypoxia or cobaltous ions. Erythropoietin is primarily produced by the kidney. Erythropoietin stimulates the synthesis of RNA, Haemoglobin and several enzymes in erythroid cells. It facilitates the entry of iron into bone marrow for haemoglobin synthesis. It increases the release of erythroid cells into circulation.

d. Promine and Retine

Certain extracts of calf thymus gland inhibit malignant growth in tumor bearing mice while other extract promote growth. The growth inhibitor is called Retine and growth promoter is called Promine. Promine and Retine is low molecular weight compound, which are normally found to be colloidial carrier.

e. Chalones

Substances, which selectively inhibit the mitotic activity of specific tissues, are called Chalones. They have been extracted from liver, kidney, epidermis and granulocytes. This mitotic inhibitor is tissue specific. The epidermal Chalone inhibit the mitotic activity of the tissue in which it is produced. Since the mitotic activity of epidermal tissue is inversely attacked by Chalones therefore Chalones are responsible for normal mitotic control in the epidermal tissue.

II. Hormones

Probably all hormones directly or indirectly influence growth by altering biochemical reactions; many of them influence the size of specific tissues and organs.

a. Somatotropin (Growth Hormone)

It is secreted by the anterior pituitary and is essential for normal metabolism in adults. Somatotropin affects many aspects of metabolism, but the two primary effect in promoting growth involve skeletal and protein metabolism. Growth hormone is species specific and is effective within the same species only. .Somatotropin stimulates the growth of epiphyseal plates of long bones and of endochondral bone in growing animals.

Low concentration causes Dwarfism
High concentration before puberty causes Gigantism
High concentration after puberty causes Acromegaly

b. Thyroxin

Thyroxin is an iodinated tyrosine compound secreted by the thyroid gland. It is synthesized in the epithelium of the thyroid follicles after iodine is transported into the gland from the circulation. In animals, normal growth and development occur only in the presence of thyroid hormones indicating that they play a permissive role in growth regulation. Since they are necessary for normal growth, thyroid hormones are classified as growth stimulators when present in optimal amount.

c. *Corticosteroids*

The adrenal cortex secrets two main corticosteroids Glucocorticoids and Mineral corticoids. The width of the epiphyseal plate decreases in growing adrenalectomized rats, indicating that adrenal corticosteroids are necessary for maintenance of bone growth in young animals. They also decrease the stimulatory effects of STH on somatomedins which is responsible for bone growth. Excessive amount of mineral corticoids are catabolic to bone and produce a negative calcium balance.

d. *Androgens*

The common observation that mature male animals are larger than females of the same age shows that sex hormones are important in growth regulation. Male and female rat grow at approximately equal rates until the beginning of sexual maturity, male then begin to grow faster and are larger than females when growth ceases in both the sexes. Androgens are more effective in stimulating skeletal growth in males than in females because estrogens, which promote epiphyseal closure, are present in smaller quantities in male. In addition to skeletal effect androgen also affect protein metabolism. The primary effect of androgens on protein metabolism is to stimulate protein synthesis.

e. Estrogen

Estrogens affect somatic growth of animals in addition to effect on sex organs. The major estrogens

Table 1. Effects of Hormone on Growth

Hormones	General effect	Effects on skeleton	Effects on Protein metabolism
Somatotro-pin	Anabolic	Stimulates growth of endochondral bone and epiphysis of long bones.	Increases N retention and protein synthesis. Thyroxin and insulin stimulate effect
Thyroxin	Anabolic	Stimulate growth of long bones, essential for the effect of STH	Stimulate protein synthesis and BMR. Deficiency or excess is catabolic.
Testosterone	Anabolic	High dose is weak stimulator of epiphyseal closure and antagonist to STH; low dose increases width of epiphysis and augments effects of STH	Increases N-retention; promotes muscles growth and development of sex characteristics.
Estrogens	Catabolic	Inhibits skeletal growth; promotes epiphyseal closure	Increases N-retention in ruminants, High doses catabolic in other animals; promotes growth of sex hormones
Glococorti-coides	Catabolic	Decreases growth of epiphysis; decreases stimulation on epiphysis by STH	Increases protein and amino acid degradation; inhibit protein synthesis in extrahepatic tissue.

produced by the ovary, placenta and to a minor extent by adrenal cortex of both sexes are Estradiol, Estrone and Estriols. Estrogen inhibits skeletal growth by decreasing growth to the long bones.

III. Feed Additives and other Growth Factors

a. Synthetic hormone like compound

The use of natural and synthetic estrogens and progesterone in animal feeding has contributed substantially to increased production, essentially in fattening animals. Implantation of a synthetic compound diethylstilboestrol (DES) increases the rate of gain by about 0.3 to 0.51 per day in big cattle. Diethylstilboestrol is effective in female, castrated males and in growing bulls, but it is more effective when rate of gain is high. Feed efficiency is generally increased by DES administration.

b. Naturally occurring growth factors in feeds

At least 50 plant materials produce estrogenic effects in animals. The amount of estrogenic substances present in the plants is important in livestock production because they sometimes produce and cause genital and mammary development. Many naturally occurring growth inhibitors are also present in livestock feed. The trypsin inhibitors of Soyabean and the gossypol content of cottonseed are the substance which causes growth inhibition.

c. Antibiotics

Antibiotics and arsenicals are used extensively as growth stimulants, especially in poultry and swine production. These substances increase growth and feed efficiency but they do not alter the mature body size of the animals. In healthy animals there is usually little improvement in gain, but antibiotics reduce the number of unthrifty slow growing animals. Antibiotics are effective in eliminating sub clinical infections and diseases and thus create a better environment for rapid growth.

d. Anthelmintics

Anthelmintics have been used as promoters of growth under practical feeding condition. Beneficial effects of anthelmintics are generally related to the control of parasitic infections. An organic phosphate 2,2 dichloro vinyl dimethyl phosphate is used to stimulate growth in fattening cattle.

e. Tranquilizers

Feeding of tranquilizers sometimes causes a slight increase in weight gain and feed efficiency in fattening beef and sheep, but there are usually no beneficial effects in other animals.

—★★★—

Chapter 4

Aging

Aging is a biological process, which leads to an increasing probability of the degenerative diseases. All animals have definite life cycle which includes birth, a period of rapid growth, puberty, a gradual cessation of growth, senescence and finally death. The pattern is rather rigidly fixed by the genetic background of the organism and can be varied only by rather drastic measures. Animals have very efficient repair mechanism, which allow them to recover from almost any type of damage or disease and yet in the absence of demonstrable malfunction, the organism degenerate and finally succumb to one of the degenerative diseases.

Anatomical Changes

In most mammals, growth virtually stops soon after puberty and body weight remains relatively constant for rest of the life span. Each animal has a preferred body size at any age and will attempt to attain this size if proper food is available. The common observations with regard to aging are

1. Decline in muscular performance

As the animal grows and increases the size and strength of the muscles, the size of the individual fibers increases but not their numbers. In senescence, the number of fiber decreases, the muscle mass tends to decrease with age, but the amount of decrease is quite variable.

2. Loss in the central nervous system

It has been estimated that the average man loses about 10,000 brain cells per day out of about 10 billion (Brody, 1955).

3. Accumulation of Collagen

Once the collagen is deposited it is very insoluble and tends to remain in the tissues for years. Many of the connective tissue cells die and are replaced by collagen fibers. These fibers tend to shrink and join in the course of time, forming a dense network. This accounts for the wrinkled appearance of the skin of old persons. In addition, the dense network tends to impede normal function.

Most of the other organs of the body tend to decrease somewhat in size as the individual ages. The liver and kidney decreases by about 20 percent in man between 25 and 70 years of age and spleen by as much as 40 per cent. The endocrine glands undergo rather profound changes depending on the function performed. Another tendency in both animal and man is to accumulate fat, as they grow older. This obesity varies greatly among different animals and even among individuals of the same species.

Physiological Changes

The most striking decrement with age is that of muscular performance. The control system of the body suffers a small linear decline with age. When placed in a cold environment an old rat takes much longer to adjust its metabolism to the changed condition to does a young one. If one injects glucose into the blood, a young animal can immediately mobilize the body defenses to bring the blood glucose levels to normal, but an older animal takes much longer. The endocrine glands undergo important changes in function during growth, development, maturation and aging. When the endocrine glands are examined anatomically and physiologically they are found to be functioning quite normally. In old age blood level of sex hormone is quite low it is not because the glands are incapable of maintaining higher level but because the functional state of the animal at that time requires that level. Reaction time becomes progressively longer with age. The changes are relatively small for simple reflexes, but for responses requiring several different senses and muscles, reaction can be very much slower in old age.

Degenerative Changes

The decline in organs function is very important in relation to aging, but the end result of aging, death is related to this decline in only a vary peripheral way if at all. In general the cause of death of older domestic animals is degenerative diseases. These comprise the various forms of cancer, the cardio vascular diseases and autoimmune diseases. The degenerative diseases can be characterized as one arising primarily from within the body itself. For some reason some of the cell of the body don't function properly and this may cause the entire machinery of the body to malfunction.

Theories of Aging

Stress Theory or Wear and Tear Theory

According to this idea, a stress of any kind destroys tissue in the individual; recovery is never complete. As the unrecovered damage from more stresses accumulates, the organism is finally unable to resist one of the degenerative diseases.

When we are young, the body's own maintenance and repair systems keep compensating for the effects of both normal and excessive wear and tear. With age the body loses its ability to repair damage caused by diet, environmental toxins, bacteria or a virus. Thus, many elderly people die of diseases that they could have resisted when they were younger.

The Neuroendocrine Theory

This theory was developed by Vladimir Dilman. According to this theory, the hypothalamus controls various chain-reactions to instruct other organs and glands to release their hormones. The hypothalamus also responds to the body hormone levels as a guide to the overall hormonal activity. However, as we grow older the hypothalamus loses its precision regulatory ability, accordingly, as we age the secretion of many hormones declines and their effectiveness is reduced due to the receptors down-grading. When we're young hormone levels tend to be high, as we age the body produces lower levels of hormones, which can have disastrous effects on our functioning. Hormones are vital for repairing and regulating our body functions, and when aging causes a drop in hormone production, it causes a decline in our body's ability to repair and regulate itself as well. Moreover, hormone production is highly interactive. The drop in production of any one hormone is likely to have a feedback effect on the whole mechanism, signaling other organs to release lower levels of other hormones which will cause other body parts to release lower levels of yet other hormones.

Thus hormone replacement therapy, a frequent component of any anti-aging treatment, helps to reset the body's hormonal clock and so can reverse or delay the effects of aging.If our hormones are being produced at youthful levels in a very real sense the cells of our bodies are stimulated to be metabolically active and thus we stay young.

Genetic Control Theory

This planned-obsolescence theory focuses on the genetic programming encoded within our DNA. We are born with a unique genetic code, a predetermined tendency to certain types of physical and mental functioning, and that genetic inheritance has a great deal to say about how quickly we age and how long we live. Each of us has a biological clock ticking away set to go off at a particular time, give or take a few years. When that clock goes off it signals our bodies first to age and then to die.

Free Radical Theory

In a conventional molecule, the electrical charge is balanced. Electrons come in pairs so that their electrical energies cancel each other out. The free radical on the other hand has an extra negative charge. This unbalanced electrical energy tends to make the free radical attach itself to other molecules as it tries to steal a matching electron to attain electrical equilibrium. In doing so, they create free radicals and extensive body damage.

Free radicals are known to attack the structure of cell membranes, which then create metabolic waste products. Such toxic accumulation interferes with the cells ability to repair and reproduce them. They disturb DNA and RNA synthesis, interfere with synthesis of protein, lower our energy levels, prevent the body from building muscle mass and destroy cellular enzymes, which are needed for vital chemical processes. This type of free-radical damage begins at birth and continues until death. In young age its effects are relatively minor since the body has extensive repair and replacement mechanisms. With age, however the accumulated effects of free-radical damage begin to take their toll. Free radicals attack collagen and elastin, the substances that keep our skin moist, smooth, flexible and elastic. These vital tissues fray and break under the assaults of free radicals, a process particularly noticeable in the face, where folds of skin and deep-cut wrinkles are testaments to the long-term effect of free-radical damage.

Substances that prevent the harmful effects of oxidation are known as antioxidants. Natural antioxidants include vitamin C, vitamin E and beta carotene.

Another substance that combats free-radical damage is known as a free-radical scavenger. Free-radical scavengers actually seek out free radicals and harmlessly bind them before they can attach themselves to other molecules and/or cause cross-linings.

The Membrane Theory of Aging

According to this theory it is the age related changes of the cells ability to transfer chemicals, heat and electrical processes that impair it. As we grow older, the cell membrane becomes fewer lipids (less watery and more solid). This impedes its efficiency to conduct normal function and in particular, there is a toxic accumulation. This cellular toxin is referred as lipofuscins and with age lipofuscins deposits become more in the brain, heart and lungs and also in the skin. An excess of lipofuscins in the body is shown as a darkening of the skin in certain areas so called "aging spot." Substances that have ability to remove lipofuscins include Centrophenoxine, DMAE and amino acid Acetyl-L-carnitine and Carnosine.

Waste Accumulation Theory

In the course of their life spans, cells produce more waste than they can properly eliminate. This waste can include various toxins which when accumulated to a certain level, can interfere with normal cell function, ultimately killing the cell. For ex. Presence of a waste product called lipofucsin leading to age pigment. The cells most commonly found to contain lipofucsin are nerve and heart muscle cells, both critical to life. Lipofucsin is formed by a complex reaction that binds fat in the cells to proteins. This waste accumulates in the cells as small granules and increases in size as a person ages.

Limited Number of Cell Divisions Theory

The number of cell divisions is directly affected by the accumulations of the cell's waste products. The more wastes accumulate over time the faster cells degenerate.

Hayflick Limit Theory

This theory of aging, suggest that the human cell is limited in the number of times it can divide. The human cells ability to divide is limited to approximately 50 times, after which they simply stop dividing and hence die. Nutrition has an effect on cells, with overfed cells dividing much faster than underfed cells. This theory indicates the need to slow down the rate of cell division if we want to live long lives.

Death Hormone Theory (DECO)

According to this we are born with roughly 12 billion of nerve cells and over a life time about 10 percent perish. This theory states that as we age the pituitary begins to release DECO (decreasing oxygen consumption hormone or "death hormone"), which inhibits the ability of cells to use thyroxin, a hormone produced by the thyroid-governing basal metabolism, the rate at which cells convert food to energy. The metabolic rate brings on and accelerates the process of aging.

Thymic-Stimulating Theory

The thymus is the master gland of the immune systems. The size of this gland reduces from 200 to 250 grams at birth to around three grams by age 60. Studies have shown that thymic factors are helpful in restoring the immune systems of children born without them as well as rejuvenating the poorly functioning immune systems of the elderly. Thymic hormones may also play a role in stimulating and controlling the production of neurotransmitters and brain and endocrine system hormones which means they may be the pacemakers of aging itself, as well as key regulators responsible for immunity.

Mitochondrial Theory

Mitochondria are the energy-producing organelles in the cells that are responsible for producing ATP, our primary source of energy. They produce cell energy by a process that leads to the formation of potentially damaging free radicals. Mitochondria are also one of the easiest targets of free-radical injury because they lack most of the defenses found in other parts of the cell.

Errors and Repairs Theory

The production of proteins and the reproduction of DNA sometimes is not carried out with accuracy. The body's DNA is so vital that natural repair processes kick in when an error is made. However, the system is incapable of making perfect repairs on these molecules every time, therefore the accumulation of these flawed molecules can cause diseases and other age changes to occur. If DNA repair processes did not exist, scientists estimate that enough damage would accumulate in cells in one year to make them nonfunctional.

Cross-Linkage Theory

The cross-linking theory of aging is also referred to as the Glycosylation theory of aging. According to this theory, it is the binding of glucose to protein that causes various problems. Once this binding has occurred the protein becomes impaired and is unable to perform as efficiently. The cross- linking of proteins may also be responsible for cardiac enlargement and the hardening of collagen, which may then lead to the increased susceptibility of a cardiac arrest. Collagen is one of the most common proteins found in the skin, tendons, ligaments, bone and cartilage. In young people there are few cross-links and the ladders are free to move up and down. The collagen stays soft and pliable. With age, however the number of cross-links increases, causing the skin to shrink and become less soft and pliable. It is thought that these cross-links increase, causing the skin to shrink and become less soft and pliable. It is thought that these cross-links begin to obstruct the passage of nutrients and waste between cells.

Cross-linking also appears to occur when older immune systems are incapable of cleaning out excess glucose molecules in the blood. These sugar molecules react with proteins causing cross-links and the formation of destructive free radicals. Scientists once thought inflexibility of the body with age was due to cross-linking of tendon, bone and muscle tissue. However, people who lead a more active life style and follow a good diet seem to inhibit or delay the cross-linking process.

Autoimmune Theory

The immune system is the most important line of defense against foreign substances that enter the body. With age, the system's ability to produce necessary antibodies that fight disease declines. In a sense, the immune system becomes self-destructive and reacts against itself. Examples of autoimmune disease are lupus, scleroderma and adult-onset diabetes.

Caloric Restriction Theory

Underfeeding particularly restriction of calorie intake increases the life span. The effect of the restricted energy intake was to slow down the growth rate and to delay the attainment of maturity. In other words, the restricted calorie intake delayed the onset of senescence by prolonging the juvenile period.

Gene Mutation Theory

Spontaneous mutation develops in the cells of the body, and since mutations are irreversible, they may build up to rather high level in due course of time. It is certain that virtually all such mutation would be deleterious to the function of the cells. If the mutation is in one of the genes controlling a vital function of the cell, the cell will die. If the mutation is in some other gene, it may change the function of the cell to make it less efficient or to make it grow in an abnormal cancerous growth. Mutations in the somatic cells of the body play a dominant role in the aging of the animals however so many other factors are also important.

The Rate of Living Theory

This theory states that we are each born with a limited amount of energy. If we use this energy slowly then our rate of aging is slowed. If the energy is consumed quickly aging is hastened. Other rate-of-living theories focus on limiting factors such as amount of oxygen inhaled or number of heartbeats spent.

Order to Disorder Theory

From the time of conception to sexual maturation, our body undergoes a system of orderliness. Most of our energy is directed to fulfill a genetically determined plan for the orderly production and arrangement of an enormous number and variety of molecules. After sexual maturation, however this energy start to diminish in efficiency. Disorder occurs in molecules in turn causing other molecules to produce errors and so on. These chaotic changes in our cells, tissues and organs causes aging. Disorderliness varies from individual and this may be the reason why our tissues and organs deteriorate at different rates.

The Telomerase Theory of Aging

A new theory of aging that holds many promising possibilities for the field of anti-aging medicine is the telomerase theory of aging. Telomeres act to maintain the integrity of chromosomes. Every time when the cells divide, telomeres are shortened. This shortening of telomeres is believed lead to cellular damage due to the inability of the cell to duplicate itself correctly. Each time a cell divides it duplicates itself a little worse than the time before, thus this eventually leads to cellular dysfunction, aging and indeed death.

Telomerase, an enzyme found only in germ cells and cancer cells, appears to repair and replace telomeres helping to re-regulate the clock that controls the life span of dividing cells. Recent research indicates that telomeres can be repaired by the introduction of the relevant genes.

An enzyme beta galactosidase found in aging cells inhibits the further cell division. Recently it has been observed that the pineal gland secretion (melatonin) that is also known as miracle hormones slows down the ageing process. It acts as powerful antioxidants.

—★★★—

Chapter 5

Bioenergetics

Definition of Bioenergetics

1. Biological study of energy transformation is known as bioenergetics.
2. Bioenergetics is the study of how cellular metabolism governs the interaction between the cells.
3. The scientific study of the use of energy by living organisms, including its conversion from one form to another.
4. The study of energy in life
5. The study of the flow and transformation of energy in and between living organisms and between living organisms and their environment.
6. A therapeutic approach that incorporates breathing, movement, body exercises, psychotherapy, and free

expression of feelings to enhance self-awareness and well-being and relieve physical and emotional tension.

7. Mind-Body therapeutic technique to help a person get back together with his body and to help him/her enjoy to the fullest degree possible the functions of the body (sexuality, breathing, moving, feeling and self-expression).Examples: Bioenergetics is a form of psychotherapy based on the work of Wilhelm Reich.
8. The shell, in Bioenergetics terms, represents defenses.
9. The scientific study of the use of energy by living organisms, including its conversion from one form to another.
10. Bioenergetics is the study of how organisms manage their energy resources.

Some Common Terms

Metabolic rate: amount of energy an animal uses per unit time, generally expressed as kcal/day.

Heterotrophs: animals that harvest chemical energy from the food they eat.

Endothermic: maintains their body temperature by body temperature its e.g. birds and mammals

Ectothermic: Maintains their body temperature by external temperature e.g. fish, amphibians, reptiles

The first law of thermodynamics is the law of conservation of energy and holds that energy can be transferred or transformed but neither created nor destroyed.

The second law of thermodynamics states that all forms of energy are quantitatively convertible to heat, that heat is the lowest energy form and that the driving force of all energy transactions is the tendency to reach the lowest energy form, i.e. heat.

Gross energy (GE) is the amount of energy manifested as heat when the feed is completely oxidized.

Digestibe energy (DE) is the proportion of the gross energy of a feed which is apparently digested. DE = GE - FE; gross energy minus energy in faeces (FE) measured in a bomb calorimeter.

Metabolizable energy (ME) is the amount of energy available to the animal for body functions (maintenance, growth, production of milk and eggs, etc.).

Heat energy (HE) is the energy produced by the total metabolism in the body. It is a sum of energy derived from oxidation of carbohydrate, fat and protein, and fermentation processes. Total heat energy includes heat produced by maintenance processes (HEm) and the so-called heat increment (HI).

Heat increment (HI) is the heat produced by the metabolism of feed and the heat produced by the conversion of the metabolized nutrients into animal products, e.g. protein and fat retained in the body, eggs, milk etc.

Retained energy (RE) is the amount of energy retained in the body and/or in milk and eggs (so called energy balance). RE = ME - HE; metabolizable energy less heat production.

Net energy (NE) is the energy of a feed which is available to the animal for different life processes;

NE = GE - (FE + UE + CH4E + HI). The deduction of the heat increment of a feed from its metabolizable energy gives the net value of a feed.

Basal metabolism which is the irreducible energy cost of maintenance during complete rest.

Special Dynamic action. an energy waste incident to food utilization.

Bioenergetics: An Introduction

1. History of heat production measurements

"La réspiration est donc une combustion" (Lavoisier, 1780)

At the end of the eighteenth century, Lavoisier showed - by means of the first biocalorimeter - that one volume of oxygen in the organism combines with carbon to form one volume of carbon dioxide, just as in combustion of carbon outside the body, and this combustion is the source of animal heat production. He also observed that the intake of nutrients increases oxygen consumption and carbon dioxide production, hence causing increased heat production in the organism. In his studies of metabolism in relation to muscular activity, Lavoisier found that the O2 intake, and thus combustion, increased with increased muscular activity. These studies brought Lavoisier to the conclusion that the combustion processes in the body are the sources of "heat production, muscular strength, as well as all other forces in the body." However, in 1794 the guillotine of the French revolution put an end to his work, and about 100 years had to pass before Lavoisier's theories were confirmed and used as a basis for calculation of energy metabolism and hence for the science of nutrition.

At the end of the nineteenth century, Rubner adduced experimental evidence in support of Lavoisier's theory that energy exchange in the body acts according to the principle of the conservation of energy. He showed that different nutrients produce the same quantity of energy on combustion in the body as outside the body when the end products are the same. Rubner's results were later confirmed by Atwater and Armsby, who showed that the quantity of energy produced by complete combustion of the nutrients outside the body is equivalent to the quantity of energy leaving the body as heat, energy excreted with the faeces and urine, and/or stored in the form of protein and fat. The results of those experiments may be described in an equation which may be considered as the fundamental equation for the science of nutrition:

Energy in feed + E. from body = E. in faeces + E. in urine + E. in methane + E. in heat + E. retained

At the beginning of the 20 century, a number of systematic studies of the energy metabolism of domestic animals were initiated. In Germany, Kellner used balance and heat production measurements to develop the starch equivalent system for evaluation of the energetic value of feedstuffs for cattle. After his death in 1912, his idea was continued in Denmark, where Møllgaard evaluated the net energy value of feed and also carried out a number of systematic measurements of the energy metabolism in dairy cows. Together with the feeding trials made by Frederiksen, these measurements form the basis of the Danish energy standards for milk production. During the last 50 years the developments in measuring techniques have resulted in a considerable knowledge about the energy expenditure in animals and humans and contributed valuable information for nutritional, environmental and medical sciences.

2. Introduction to bioenergetics

The word energy originates from Greek and means "in work" (*en ergon*). Energy is the capacity to do work and can be measured only in its transformation from one form to another.

Originally, all measurements of energy transactions made by animal nutritionists were expressed in terms of kilocalories. The calorie (cal) is defined as the amount of energy required to raise temperature of 1 g of water by 1 C, measured from 14.5 to 15.5 C. Since this unit is too small for most convenient use in nutrition, it has been replaced by the kilocalorie (kcal). Although the calorie is the basic unit of heat energy, the joule was adopted in the International System of Units as preferred unit for expression of electrical, mechanical, and chemical energy (Table 2). The joule is defined in terms of the International Metric System as one kg-m2/sec2 or 107 erg. Because all measurements of energy are defined in terms of the fundamental metric units of mass (kg), distance (m), and time (sec), joules are readily converted to calories and vice versa. Measured in terms of joule 1 kcal = 4.185 kilojoules (kJ). The heat-technical calculations are based on the unit of watt with the use of a heat production unit (hpu) defined as 1 hpu = 1000 W. Since 1 W = 0.860 kcal/hour or 3.6 kJ/hour, 1 hpu = 3600 kJ/hour.

Table 2. Energy units used in biological and technical calculations

Biological calculations :		
1 cal	=	4.1855 J
1000 cal = 1 kcal (kilocalorie)	=	4.1855 kJ (kilojoule)
1000kcal = 1 Mcal (megacalorie)	=	4.1855MJ (megajoule)
Technical calculations :		
1 W (watt) = 0.860 kcal/hour	=	3.6 kJ/hour
1000 W = 1 kW (kilowatt) = 1 hpu	=	3600 kJ/hour

There are two categories of energy: potential and kinetic. Potential energy is energy of position or bound energy which can be released to produce kinetic energy or to do work.

Kinetic energy is energy of movement. In the living organism the energy of random molecular motion is heat energy. Energy supplied to animals in chemical form can be transformed to heat when utilized in support of life processes, stored in chemical form as in growth, transferred in chemical form to a second animal as in pregnancy and lactation, or transferred to surroundings as in work. Bioenergetics of animals and man is primarily based on the two laws of thermodynamics.

> The first law of thermodynamics is the law of conservation of energy and holds that energy can be transferred or transformed but neither created nor destroyed.

> The second law of thermodynamics states that all forms of energy are quantitatively convertible to heat, that heat is the lowest energy form and that the driving force of all energy transactions is the tendency to reach the lowest energy form, i.e. heat.

The 2nd law can also be stated as the tendency for energy with a high degree of orderliness to be converted to energy with a lower degree of orderliness. The degree in which the total energy of a system is uniformly distributed (randomness) and thus unavailable to do work is expressed by the term entropy. In any isolated system, entropy tends to increase because randomness is more probable than is orderliness. The more disordered or random a system becomes the more entropy it has. Some molecules have a greater order than others and so they have lower entropy. For example proteins are highly ordered but upon denaturation they change to a much more random structure, hence the increase in entropy during denaturation is considerable. Generally solids are more ordered than liquids which are more ordered than gases. According to the 2nd law, any change in the total energy content of a system (e.g., the heat of combustion in a biological oxidation) results in a change in both free energy and the entropy of the system. Since only the former can be utilized to perform work of any kind, energy-yielding reactions invariably have an efficiency of less than 100 %.

It can be also stated that in accordance with the 1st law the total energy of the universe remains constant, and in accordance with the 2nd law the entropy of universe always increases. Both laws of thermodynamics emphasize that all forms of energy (in living organism) can be measured by heat. Therefore determination of heat production is theoretically sound measure of metabolism; it is also a convenient measure.

The total energy content of an isolated system, for example of an animal, cannot be measured, but it is possible to measure the energy changes - transfer and/ or transformation - which occur within the system. The difference between the energy content of the substrates and that of products is called the heat of reaction - enthalpy of change (H). If heat is gained from the environment H is positive (exothermic reaction); if it is lost the enthalpy is negative (endothermic reaction). For example if glucose is oxidized in a bomb calorimeter, 2.8 MJ of heat are liberated per mole of glucose oxidized, the H is then -2.8 MJ. Provided one has a set of heats of formation of compounds, then the enthalpy of any reaction, whether exothermic or endothermic, can be calculated as the difference between the heats of formation of products and reactants. It is most important to note that the value of H is independent of the chemical pathway followed by the reaction, that is, it is independent of the mechanism of the reaction. Whether glucose is oxidized by burning in a calorimetric bomb in the laboratory or by the action of glycolysis in the body, H is the same. Although the enthalpy change from a specified substrate to the known product is the same whatever is the mechanism of conversion, the amount of work done - amount of useful energy - will depend on the mechanism.

The amount of useful energy- free energy - 'orderly energy' i.e. capable of doing work which is liberated in a chemical reaction, cannot be determined from the heat of reaction. An other factor must be calculated; the change in free energy (G) which is a difference between free energy of reactants minus free energy of products. This change for a reaction occurring reversibly at constant temperature and pressure is defined as: G = H - T S, where T is the absolute temperature, S is the change in entropy. If G is negative, the reaction can occur spontaneously, if G is positive, it cannot.

Potential energy derived from the metabolism of foods is stored in living organism as "high energy compounds". The breakdown of these compounds liberates free energy, which can accomplish work, together with heat energy, which cannot do work, and is thus a measure of the energetic inefficiency of the reaction. The energy liberated by chemical reactions in the cell is used to drive other reactions. These require some methods for coupling the reactions that liberate energy to those that use it, and the first step involves storing the energy derived from combustion of foodstuffs in certain high energy compounds.

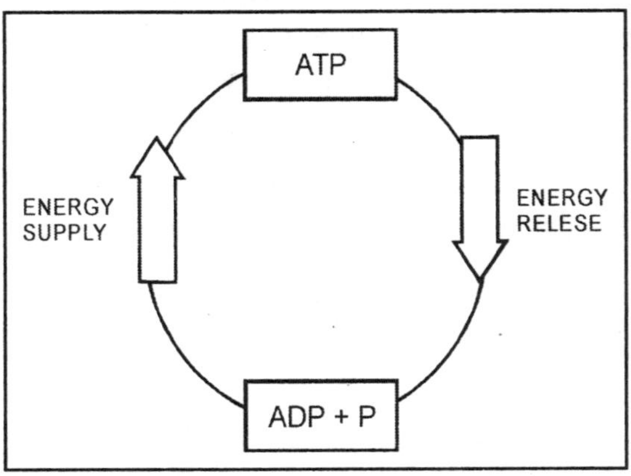

The most important of the high energy compounds is adenosine triphosphate (ATP). In a simple way, we may say that a cell is supplied with energy stored in ATP, while the energy derived from oxidation is used to resynthesize ATP. The free energy released in the hydrolysis of ATP to ADP (adenosine diphosphate) under standard conditions (standard free energy measured when 1 mole of reactant is converted to 1 mole of product at 37oC and pH=7) is generally accepted to be about -30 kJ per mole. The standard G of the reaction ADP->AMP (adenosine monophosphate) or AMP->phosphate is slightly lower, about 27 kJ/mole. However, under the conditions of pH and temperature in the body, and because most of the ATP and ADP in the intact cells is presented as Mg++ complex, which has the effect of shifting the equilibrium of ATP hydrolysis, the standard free energy change of ATP hydrolysis is much higher. On average the amount of energy that is available from the phosphate bond of ATP is 52 kJ/mole. Directly measured standard free energy of hydrolysis of phosphate compounds is demonstrated in Table 3.

Table 3. Standard free energy of hydrolysis.

Substrate	**Standard + G, kJ/mole**
phosphoenolpyruvate	- 62
1, 3-diphosphoglycerate	- 50
phosphocreatine	- 42
acetyl phosphate	- 42
ATP	- 30
Glucose 1-phosphate	- 21
Fructose 6-phosphate	- 16
Glucose 6-phosphate	- 14
3-phosphoglycerate	- 10
glycerol 3-phosphate	- 9

Although the standard free energy of ATP hydrolysis occupies an intermediate position, the role of ATP is unique. In comparison with other phosphates, ATP is the only phosphate which serves as a carrier of chemical energy between donors and low-energy phosphate acceptors. ATP is a common intermediate in both energy-delivering and energy-requiring reactions of the cell, and it is the only form of chemical energy that can be converted into all others forms of energy used by living organism. It therefore fills a role analogous to that of money - "energy currency of the cell". The main weakness of the analogy is that money, unlike ATP, can be accumulated for later use.

Energy-yielding Processes (Catabolism)

The animals require energy requirement for maintenance and production (meat, milk, foetus, eggs, wool, work). The energy is provided by degradation of the nutrients absorbed by the animal. Under normal conditions the energy is produced by oxidation processes in the three main groups of nutrients: carbohydrates, fat and protein according to the following formula:

Nutrient + O2 CO2 + H2O + energy

Catalysts (Intercellular Enzyme Systems)

The oxidation proceeds through a series of processes, in which the degradation from one stage to the next is catalyzed by intercellular enzyme systems, such as the dehydrogenases, which liberate hydrogen, and the oxydases, which activate oxygen. Part of the energy resulting from oxidation of the organic substances in the body are transferred to energy-rich phosphorus compounds (mostly ATP), from which they are later mobilized and utilized for syntheses and other life processes (work, intake and transport of food, absorption of nutrients etc.).

Oxidation of a typical carbohydrate, fat and protein can be summarized in the following stoichiometric equations:

Glucose

1 g glucose + 0.747 liter O_2 ->0.747 liter CO_2 + 0.60g H_2O - 15.65 kJ

1 mole $C_6H_{12}O_6$ + 6 O_2 -> 6 CO_2 + 6 H_2O - Energy

180 g 134.5 liter 134.5 liter 108 g 2817 kJ

Tripalmitate

1 g trip. + 2.011 liter O_2 ->1.416 liter CO_2 + 1.09 g H_2O - 39.8 kJ

1 mole $C_{51}H_{98}O_6$ + 72.5 O_2 -> 51 CO_2 + 49 H_2O - Energy

807 g 1625 liter 1143 liter 883 g 32 MJ

Meat Protein

for mammals

1 g protein + 0.992 l O_2 ->0.848 l CO_2 + 0.38 g H_2O + 0.332 g urea - 18.4 kJ

for birds

1 g protein + 0.893 l O_2 ->0.649 l CO_2 + 0.49 g H_2O + 0.498 g uric acid - 17.8 kJ

Oxidation of Carbohydrates

Carbohydrates are broken down in all cells, especially in the liver cells, where glucose is synthesised to glycogen which is stored in the liver and the muscles. Under catabolic conditions the glycogen is hydrolysed to CO2 and H2O while taking up O2.

The complete oxidation of glucose in the body may be described as follows:

C6H12O6 + 6 O2 = 6 CO2 + 6 H2O - 38 mol ATP

In the combustion of 1 mole of glucose, 6 moles O2 are used, 6 moles CO2 are produced, and 38 moles ATP (adenosine triphosphate) are liberated, which then act as energy donors for the anabolic; energy-consuming processes.

Glucose as energy source	ATP, moles
1 mole glucose to 2 moles of pyruvate	8
2 moles pyruvate to CO2 and H2O	30
Total per 1 mole glucose	**38**

Assuming that each mole of ATP contains 52 kJ, 1 mole of glucose will produce 38 x 52 = 1976 kJ in the body. Since the total energy content determined by combustion in the bomb calorimeter, i.e. outside the body, is 2817 kJ/mole, this will be equivalent to an efficiency of 70%, (1976 kJ / 2817 kJ). This shows that only 70% of the total energy of the glucose is available for the anabolic processes, while the rest is lost as heat.

The release of energy in a utilizable form from glycogen necessitates its breakdown to glucose which is then oxidized as previously described. The net effect of glycogen breakdown is the production of glucose-1-phosphate plus a little glucose. Glucose-1-phosphate and the residual glucose is converted to glucose-6-phosphate which enters pentose phosphate pathway. The production of glucose-6-phosphate does not involve expenditure of ATP so that ATP provided from glycogen is slightly higher that it is from glucose (Table 4).

Not all tissues obtain energy from aerobic processes. Erythrocytes, white blood cells, kidney, tissue of eye and cancer cells gain most of their energy from the anaerobic conversion of glucose to lactate. According to the generally accepted view, metabolism is aerobic at rest and during moderate work. At some level of work, aerobic metabolism becomes inadequate to supply energy, and energy deficit is supplied from anaerobic metabolism. Under anaerobic conditions the formation of 2 moles of lactic acid from 1 mole of glucose is associated with the net formation of only two moles of ATP. The anaerobic breakdown of 1 molecule of glucose to 2 molecules of lactic acid is called anaerobic glycolysis. Two molecules of ATP are utilized in the early stages of glycolysis - in the phosphorylation of glucose and fructose 6- phosphate. At later stages 4 moles of ATP are formed, so that there is a net yield of 2 moles of ATP for each mole of glucose utilized;

$$\text{Glucose} + 2\ \text{P} + 2\ \text{ADP} \rightarrow 2\ \text{lactic acid} + 2\ \text{ATP}.$$

Anaerobic glycolysis can be reversed when oxygen becomes available. If the oxygen supply is restored the lactic acid disappears, a portion is oxidized to CO2 and water and the reminder reconverted to glycogen.

Table 4. Yield of ATP from the catabolism of carbohydrates

Substrate	Conditions	ATP, moles/ substrate, mole
glucose	aerobic, completer oxidation	38
glucose	anaerobic, into lactate	2
glycogen	aerobic, complete oxidation	39
glycogen	anaerobic, into lactate	3

Oxidation of Fat

The majority of the cells of the body are able to oxidize and convert lipids and to use fat as a source of energy. The major part of the energy derived from fat is provided by the fatty acids. Fatty acids are made available from dietary lipids and from triacylglyceroles stored in the body when glucose is unable to provide sufficient energy. The breakdown of the lipids starts by hydrolytic processes in the intestine, where the fatty acids are split off from their esters with the glycerol part. The absorbed glycerol is oxidized like a carbohydrate via the glycolysis and the tricarboxylic acid cycle, in which process 1 mole of glycerol yields 21 moles of ATP. The fatty acids are broken down to acetyl-COA, and then to CO2 and H2O. Taking tripalmitin as a model; In the cell 408 moles ATP are formed from each mole of tripalmitin. The energy stored per mole ATP is 52 kJ, corresponding to 21216 kJ/mole tripalmitin oxidised in the body. The energy yield of palmitic acid in the bomb calorimeter is 9981 kJ/mole, 1 mole glycerol yields 1655 kJ, hence the total energy from 1 mole of tripalmitin (3 x 9981)+1655 = 31598 kJ. Consequently, the efficiency of the oxidation is then 21216/31598 = 0.75. The remaining 25% is heat loss.

In ruminant animals considerable amounts of propionic acid, butyric acid and acetic acid are produced by microbial breakdown of carbohydrates in the rumen. The intermediary products of carbohydrate breakdown in the rumen are oligosaccharides and simple sugars. These are unstable towards rumen microorganisms and, in majority, broken down to short chain fatty acids (SCFA) also called volatile fatty acids (VFA). First, the monosaccharides are converted to pyruvate and subsequently to SCFA. The conversion of sugar to pyruvate and formation of SCFA generates ATP which mainly serves as the energy source for synthesis of microbial protein. The conversion of 1 mole hexoses provides 4 moles ATP, but for pentoses the process is less efficient contributing 1.67 moles ATP. SCFA produced in the rumen pass to the tissues where they partly serve for synthesis of fat and/or glucose and partly undergo oxidation. The efficiency of oxidation of SCFA is about 60 % (Table 5).

Oxidation of Protein

Food proteins are hydrolysed to amino acids in the intestine, where they are absorbed and used to build up proteins or are oxidized. Amino acids not used for synthesis are deaminated resulting in the liberation of ammonia and the conversion of the amino acids to the corresponding keto acids. These can be used for glucose synthesis or lipogenesis and oxidized for energy supply.

The final product of amino acids degradation is acetyl-CoA and, depending on the nature of amino acids, glucose (from glucogenic amino acids) or ketone bodies (from ketogenic amino acids). The major glucogenic amino acids are alanine, glutamate and valine, while leucine is ketogenic. The other amino acids may be partly glucogenic and partly ketogenic.

Most of the ammonia arising from the degradation of amino acids is excreted as urea. Urea formation requires energy, a total of 4 moles of ATP is needed per 1 mole of urea formed. Further, about 1 mole of ATP is used during the excretion of urea. Therefore, in assessing the efficiency of amino acids oxidation, the energy required for urea synthesis must be considered in the calculations. The high energy requirement for urea synthesis and excretion may be one of the main reasons for the great heat output observed after ingestion of protein in excess of the requirement.

Taking aspartate as a model; In th cell 16 moles of ATP are formed from each mole of aspartate. The energy stored per mole ATP is 52 kJ, corresponding to 16 x 52 = 832 kJ/mole aspartate oxidized in the body. The energy yield of aspartate in the bomb calorimeter is 1568 kJ/mole and consequently the efficiency of the oxidation is 832/1568= 0.53.

The data in Table 5 summarized the estimates of heat energy equivalents of high energy phosphate bonds formed during the oxidation of several major energy sources.

Table 5. The enthalpies of combustion (Hc), yield of ATP on the complete oxidation, the energy required for formation of 1 mole ATP (Hc/ATP) and the efficiency of oxidation

Energy source	+Hc kJ/mole	ATP moles	+ Hc/ATP kJ/mole	Efficiency pct
Glucose	2817	38	74	70
Glycogen	2839	39	73	71
Acetate	875	10	88	59
Butyrate	2184	25	87	59
Propionate	1527	17	90	58
Palmitate	9981	129	77	67
Sterate	11285	146	77	67
Glycerol	1655	21	79	66
Alanine	1625	12.5	130	40
Glutamate	3213	23	140	37
Lysine	3682	31	119	44

Energy-Requiring Processes (Anabolism)

Most of the energy transferred to ATP via the oxidative processes is later liberated and used in the anabolic processes. Quantitatively the anabolism mainly concern the formation of carbohydrate (glycogen in the liver and the muscles; lactose in milk), of fat (tissue, milk, foetus, eggs) and protein (tissue, enzymes, hormones, milk, foetus, eggs, hair). Genetically the urge of forming protein is quantitatively much more dominating than of forming fat. In young growing animals the body gain may comprise 25% protein, 70% water and 5% fat, which is mostly as structural fat. As the animals approach maturity the preference toward protein retention decreases and that toward fat accretion increases; in full grown animals a body weight change may consist of less than 5% protein and more than 80% fat. By fat synthesis often a too great energy supply can be stored as a reserve for later mobilization in case of insufficient energy supply from food. Apart from the long-term regulation of energy supply by means of reserve fat accretion and mobilization, there is a short-term regulation of energy supply. Directly after the intake of food the liver stores a large part of carbohydrate as glycogen, and later on these stores are broken down to glucose.

Glucose is a substrate for production of glycogen stores and lactose. In ruminants storing and use of glycogen are less developed than in monogastric animals. Apart from absorbed glucose, glucose is synthesized from gluconeogenic substances. Lactic acid, amino acids, glycerol, and - in ruminants - propionic acid are the main gluconeogenic substances. In lactating cows, 55% of the gluconeogenesis is from propionic acid, 20% from lactic acid, 15% from amino acids, and 10% from glycerol.

Glycogen is a complex polysaccharide made up of glucose, and has the ability to add on further glucose units. It is a readily mobilized storage form of glucose. The formation of glucogen is a fairly cheap process, totally, slightly more than 1 mole ATP per mole glucose 1- phosphate is used for the storage in glycogen. Hence, 1 mole ATP is used to phosphorylate glucose to 1 mole glucose 1-phosphate and one high-energy phosphate bond is spent in incorporating glucose 6-phosphate into glycogen.

Fatty acids synthesized in the body and fatty acids absorbed from the intestine are concerned in the esterification of fatty acids to triglycerides and certain phosphatides in depot fat. There are two systems of fatty acid synthesis. The first is a cytoplasmic system resulting in the production of palmitic acid from acetyl-CoA. The second is a mitochondrial system resulting in the elongation of the existing fatty acids by two-carbon addition by means of acetyl-CoA. Most unsaturated fatty acids are formed by dehydration of the corresponding saturated fatty acids, except the essential fatty acids, which cannot be produced by the animal, but must be supplied with the feed. Three moles of fatty acid and 1 mole of glycerol are used for complete synthesis of a triglyceride. The glycerol part is formed from glucose by reduction of dihydroxyacetone phosphate to glycerol.

Proteins are synthesized from amino acids, which become available either as the end products of digestion or as a result of synthetic processes within the body. In the first step of the protein synthesis the amino acids are linked to form peptides in a genetically determined order. Then the peptides are polymerised and folded to protein by means of different linkages. During all stages of protein synthesis energy is provided by hydrolysis of ATP and guanosine triphosphate (GTP), the total number of moles of ATP and GTP needed to synthesize one peptide bond is 4. The efficiency of 87% (Table 6) is about 30% higher than the measured energy required for protein retention. The main cause of the large difference between energy requirement for synthesis and energy requirement for retention is the protein turnover (synthesis/ degradation). Isotope studies have revealed that the ratio of protein synthesis/retention is not constant; depending on the age of the animals, 2-5 times more protein is synthesized than is being stored. Assuming that the average ratio of synthesized/retained protein is 3/1, the energy efficiency for protein accretion will be only 69% as against 87%.

The data in the Table 5 summarize the estimates of theoretical biochemical efficiencies for deposition of protein, fat and carbohydrate calculated as enthalpy of combustion of the product divided with enthalpy of combustion of the substrate + energy for synthesis from ATP.

Table 6. Theoretical stoichiometric efficiencies with which the energy of nutrients is employed in synthesis

Substrate	Product	Efficiency pct.	Heat increment pct.
Glucose	Body glycogen	95	5
	Body fat	82	18
	Milk lactose	94	6
Lipid	Body fat	96	4
Protein	Body protein	87	13
	Body fat	66	33
	Milk protein	86	14
Propionate	Milk lactose	78	22
Acetate	Milk fat	76	24

Summary

Energy is the capacity to do work and can be measured only in its transformation from one form to another. Energy supplied to animals in chemical form can be transformed to heat, stored in chemical form as in growth, transferred in chemical form to a second animal as in pregnancy, or transferred to surroundings as in work. Bioenergetics of animals and man is based on the laws of energy conservation and increasing entropy. Energy derived from the metabolism of foods is stored in living organism as high energy compounds. The most important of the high energy compounds is ATP. ATP serves as a carrier of chemical energy between energy donors and acceptors, it is a common intermediate in both energy-delivering and energy-requiring reactions of the cell.

The energy in the body is provided by oxidation processes of carbohydrate, fat and protein. The oxidation processes in the body break down the nutrients to carbon dioxide and water, and part of the chemical energy is transferred to the energy-rich phosphorus compounds. The remaining part of the energy is used to maintain the constant temperature of the body and is given off as heat.

The greater part of the energy of the energy-rich compounds is liberated for synthesis of the tissues of the body, while the rest is used for the physical processes such as activity, feed intake and transport, absorption of nutrients from the intestinal tract, blood circulation, maintenance of fluid balance, ion transport etc. and is given off as heat (Fig. 6).

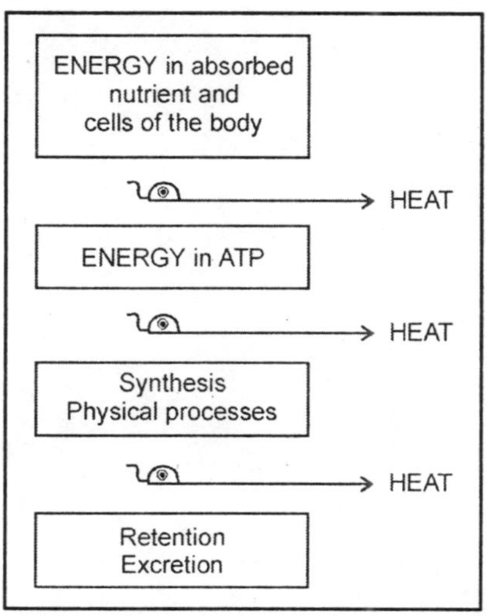

Fig. 6. Energy-producing and energy-requiring process in the body.

—★★★—

Chapter 6

Methods for Quantitative Determination of Heat Production

The metabolic processes responsible for the energy supply to the body can be determined by measuring the heat production of the animal. The quantity of energy in the oxidative processes of carbohydrates, fat, protein and short chain fatty acids can be measured as heat. In the previous chapter we calculated that the animal utilizes about 70% of carbohydrate and fat energy, and 40% of protein energy for energy-requiring processes, while the rest is lost as heat. Part of the energy transferred to the energy-rich phosphorus compounds and later hydrolyzed will also, sooner or later, be converted to heat. This means that all the metabolic processes can be quantified on the basis of the heat production of the animal.

Heat production can be measured by means of various calorimetric methods, either directly in a calorimeter or indirectly by measuring the gas exchange from the animal.

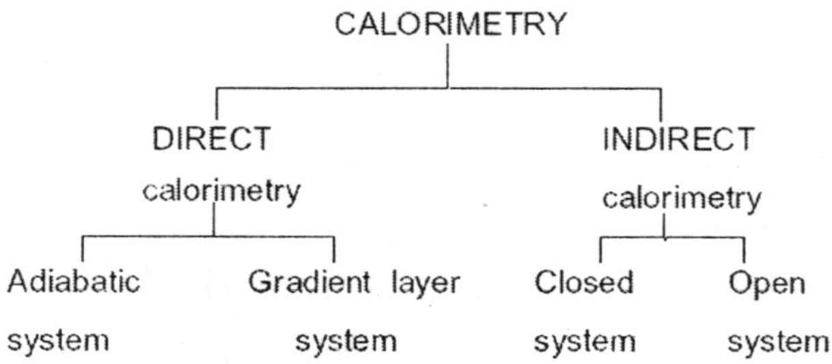

Direct Calorimetry

An animal calorimeter is an apparatus for direct measurement of the total heat production of an animal. Lavoisier and Laplace (1780) were the first to measure animal heat production. The animal was enclosed in a chamber surrounded by ice, and the amount of ice melted within a given period of time, multiplied by the melting heat of the ice (330 kJ/kg), was recorded as the measure of the heat given off by the animal. Provided that all the heat produced was free heat, it would be equivalent to the total heat production of the animal.

Towards the end of the 19th century, Atwater and Ross constructed a water calorimeter (adiabatic principle) in which the heat production was measured according to the same principle as in a bomb calorimeter, but where the chamber was placed in an insulated tank of water in which the amount and temperature of water was recorded.

Fifty years later, the technique of measuring heat flow was so advanced that Bentzinger and Kitzinger

(1949) were able to elaborate theories about devising a calorimeter according to the gradient layer principle. Pullar in Scotland, managed to construct such a calorimeter, which measures the quantity of heat electrically by means of thermoelements coiled around metal strips that completely cover the surface of the chamber. When the junction of the two metal wires of a thermoelement is exposed to radiation heat, a potential difference occurs, which is measured with a galvanometer, and the total data are recorded. The latter type of calorimeter is not used a great deal for domestic animal experiments, because the system has a number of drawbacks:

1. The animal has little freedom of movement, since in order to limit the number of metal strips the chamber must be as small as possible.
2. The animal has no contact to its surroundings, because to prevent heat loss there are no windows in the chamber.
3. Feed and water have to be fed into the chamber at a constant temperature, which is technically difficult.
4. The incoming air should have exactly the same temperature as the outgoing air, which requires a very accurate air-conditioning system.

Indirect Calorimetry

The total amount of energy produced by an animal in the form of heat can be calculated from measurements of gas exchange (O2-consumption and CO2-CH4-production) combined with measurements of nitrogen excretion in urine (RQ-method) or combined with measurements of carbon-nitrogen balances (CN-method). Calculation of heat production in accordance with the two methods will be described later (Paragraph IV.2). Both methods require accurate, quantitative determination of the gas exchange, which can be achieved by means of respiration units either of co-called closed-air-circuit or open-air-circuit system.

Closed-Air-Circuit System

In this system (Fig. 7) the same amount of air is circulated from the chamber through carbon dioxide and water absorbers and back to the chamber, and the total CO2 production is determined. The oxygen used by the animal is replaced by a metered supply of pure oxygen. The amount of the produced CO2 is found directly by weighing the absorbers before and after the experiment and calculating the difference. The total oxygen consumption of the animal is determined by volumetric measurement of the oxygen supplied. Corrections must be made for the composition of the air in the respiration chamber before and after the experiment.

However, the system has several disadvantages:

1. The animal has little freedom of movement, because the chamber has to be as small as possible in order to reduce chamber correction.
2. The animal has no contact with its surroundings, since normally the chamber has no windows to reduce the risk of leakage.

3. Since the air is pumped through the chamber, it must be completely airtight and under constant supervision.
4. Continuous recording of the CO_2 production is not possible.
5. The CH_4 produced by ruminants will accumulate in the system, as continuous absorption of CH_4 is not possible.

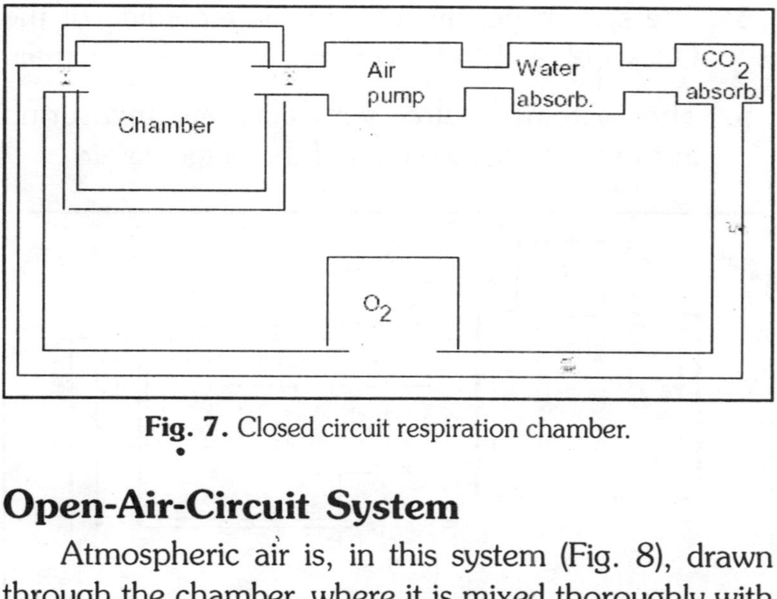

Fig. 7. Closed circuit respiration chamber.

Open-Air-Circuit System

Atmospheric air is, in this system (Fig. 8), drawn through the chamber, where it is mixed thoroughly with the air expired by the animal. The total volume of the mixed air is measured continuously by a flowmeter and its composition of O_2, CO_2 and CH_4 is measured in aliquot samples by a gas analyser. The animals' consumption of O_2 and production of CO_2 and CH_4 can be calculated based on the volume and composition of the ingoing air compared with the outgoing air. In this system:

1. The animals have reasonable freedom of movement.
2. The animals have some contact with the surroundings, because the chambers have several windows.
3. Since the system works with lower pressure than the surrounding, there is no risk of loss of chamber air, so the chambers need not be completely airtight.
4. The system permits continuous recording of the O2, CO2, and CH4 contents in the outgoing air.
5. The system requires very accurate measuring instruments for recording of the outgoing air.

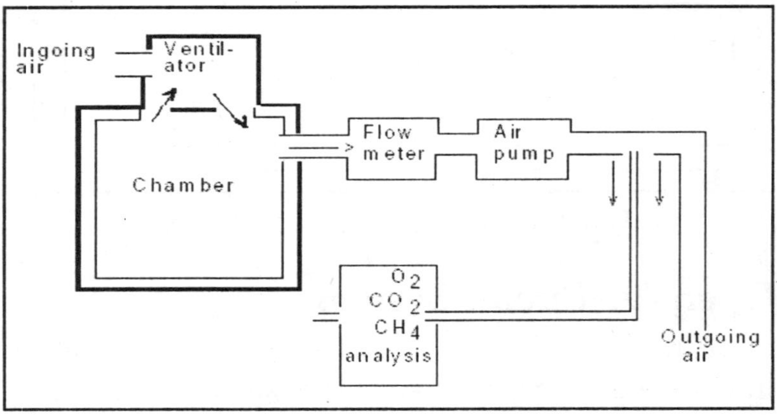

Fig. 8. Open circuit respiration chamber.

Open-Air-Circuit Units in Denmark

At the National Institute of Animal Science, Copenhagen, a respiration unit for lactating cows was already in 1915/17 constructed by Møllgaard and later (1923/33) a unit for pigs was built. The development of automatic measuring techniques after 1945 allowed Thorbek to construct more modern units for pigs (1959/61) and for cattle (1969) at the Institute. Later (1978) units for smaller animals (rats, rabbits, mink, poultry and piglets) were built by Chwalibog. At the Royal Veterinary and Agricultural University, Copenhagen, a respiration unit for measuring energy metabolism of animals from 1 to 250 kg body weight was built recently. At present time there are also respiration units for humans at the Research Department of Human Nutrition and units for small and large animals at the Danish Institute of Agricultural Sciences in Foulum.

The units mentioned, are of the open-air-circuit system and are equipped with automatic controlled systems for keepings constant air temperature and humidity. Furthermore the chambers are provided with a combined feeding and safety device, equipment for collecting faeces and urine, as well as with a milking machine in the cattle unit. By means of a gas-tight ventilator, fresh air is drawn into the chamber where it is thoroughly mixed with the chamber air. Temperature and humidity in the chamber is maintained by the air-condition system. The mixed air is drawn out of the chamber via a measuring system (differential pressure system) which records the exact volume of air leaving the chamber. When there is a change in the rate of air passage, and hence in the differential pressure, the voltage and the signals to the electric integrator and to a computer will change correspondingly. When the flow of air has passed the ventilator, it is divided so that the greater part is discharged, and a smaller sample (10%) is passed through an analysing device for determination of the composition of the air. The CO2 and CH4 concentrations are measured by means of the infrared absorption technique on the basis of the thermal conductivity of CO2 and CH4. The O2 concentration is measured on the basis of the paramagnetic principle. The electronic measuring signals from the air analyser pass to the recording unit, which graphically records the concentrations of CO2, CH4 and O2, and to the computer, where all the data are collected continuously for further calculation of the gas exchange by the animal and hence its heat production.

The results of respiratory exchange and heat production of different species obtained in Danish respiratory trials are shown in Table 7.

Table 7. Examples of respiratory exchange and heat production of different species

Spicies	**Weight kg**	**CO_2 1/day**	**O_2**	**Heat Production**
Man	70	360	500	9850
Calves	100	960	990	18300
	250	2150	2280	40800
Cows (Jersey)	350	3200	3100	65000
Pigs	5.0	110	118	2460
	20.0	400	420	8750
	50.0	720	660	14150
	90.0	1080	920	20160
Mink	1.0	31	38	770
	2.0	50	65	1270
Poultry	0.25	13	15	260
	1.00	31	36	630
	2.00	44	48	900
Rats	0.10	7.2	7.7	120
	0.20	1.3	10.3	216

Calculation of Quantitative Nutrient and Energy Metabolism

The RQ Method

According to the principle of conservation of energy nutrients such as carbohydrate, fat, and protein will produce the same quantity of energy on complete combustion in the body compared to outside the body. The heat loss depends on the type of nutrient broken down and it is therefore, necessary to know the quantities of carbohydrate, fat and protein broken down. These can be approximately calculated when the oxygen intake and amounts of carbon dioxide and nitrogen excreted with the urine are known.

Since carbohydrates and fat in the body are oxidized to carbon dioxide and water, the amount of energy released can be calculated by multiplying the amount of oxidized carbohydrate and fat by the mean combustion heat of these substances outside the body (cf. Table 8).

Oxidation of amino acids is more complicated, because the oxidation of protein produces more heat in a bomb calorimeter than in the animal body. The reason is that some of the nitrogen-containing compounds (urea, allantoin etc.) that are excreted through the kidneys contain energy. The amount of total energy of the nitrogen-containing substances lost in this process may vary considerably according to the composition of the protein and the end products of deamination. It is generally assumed that the energy loss through N-containing waste products in mammals is 21% of the combustion heat of the proteins. Since the combustion heat is 23.9 kJ/g protein, approx. 5.5 kJ will be lost during oxidation in the body, while only 18.4 kJ/g protein will be released as oxidation heat. The excretion of the N containing waste in birds differs from that of mammals in that birds excrete N in the form of uric acid, with the result that only 17.8 kJ/g protein is released on combustion of 1 g protein in the body.

The heat production of animals is often determined by the respiratory quotient (RQ), which shows the ratio of carbon dioxide produced to the amount of oxygen used. The calculations are based on the fact that the RQ value depends on the type of digested nutrient. The internationally accepted factors for calculation of energy metabolism are shown in the tables below. The physical values of the various gases are shown in Table 8, while the empirical constants for oxidation in the animal body are tabulated in Table 9. Oxidation of 1 mole of glucose will give an RQ of 1.00, since the animal uses 6 moles O2 and produces 6 moles CO2 (cf. Paragraph 3.1), that is:

RQ = 6 moles CO2 / 6 moles O2 = 1

Calculation of the total heat production by means of the RQ method requires previous determination of

Table 8. Physical values of different gases

	Mole weight g	**Litre weight g**	**Carbon g/100g**	**Carbon g/litew**	**Compustion value kJ/litre**
O_2	31.998	1.429	-	-	-
CO_2	44.009	1.964	27.29	0.5360	-
CH_4	16.043	0.716	74.87	0.5360	39.6
H_2	2.016	0.090	-	-	12.8
N_2	28.014	1.250	-	-	-

Table 9. Constants for protein, fat and carbohydrate (per g substance) for oxidation in the animal body

	Carbon %	Used in Oxidation		Produced by Oxidation		Oxidation heat	RQ mean kJ/g[2]
Protein1 (16% N)	52.00	1.366	0.957	1.520	0.774	18.42	0.809
Fat	76.70	2.875	2.013	2.810	1.431	39.76	0.711
Starch & glycogen	44.45	1.184	0.829	1.629	0.829	17.58	1.000
Saccharose	42.11	1.122	0.786	1.543	0.786	16.74	1.000
Glucose	40.00	1.066	0.746	1.466	0.746	15.65	1.000

1 Protein combustion value (in a calorimeter) = 23.86 kJ/g.

2 Heat of oxidation for fat and carbohydrate is equivalent to combustion value.

the heat produced during oxidation of protein, which can be done on the basis of the output of N in the urine. The oxidation of proteins begins with deamination. The ammonia produced is excreted into the urine in the form of urea or other N-containing compounds, e.g. uric acid, allantoin, ammonia, and hippuric acid. The N-part of these compounds originates from decomposition of proteins. Therefore, with considerable approximation, the N-excretion in the urine may be used as a measure of the amount of protein oxidized. Next, the heat produced during the oxidation of carbohydrates and fat (N free substances) is calculated on the basis of RQ values.

Using the constants for protein, fat, and carbohydrate oxidation in the body (Table 9) the calculations are carried out as follows:

1) The heat production from oxidation of protein (HEprot) is calculated as follows:

   Oxidized. prot., g = N excreted in urine, g x 6.25

   HEprot,kJ = oxid.prot., g x 18.42 kJ.

2) Oxygen used (O_2,prot) and carbon dioxide produced (CO_2,prot) during oxidation of protein: During oxidation of protein in the body 0.957 l O_2 is used and 0.774 l CO_2 is produced per g protein, which results in an RQ of 0.809. From this value, the O_2 consumption and CO_2 production are calculated by the following equations:

O_2,prot/litre = oxidized prot., g x 0.957

CO_2,prot/litre = oxidized prot., g x 0.774.

3) Oxygen consumption (O_2,N-free) and carbon dioxide production (CO_2,N-free) during the oxidation of nitrogen free (N-free) substances, ie. carbohydrate+fat, is found by:

O_2,N-free/litre = Total O_2/litre - O_2,prot/litre

CO_2,N-free/litre = Total CO_2/litre - CO_2,prot/litre.

This gives the RQ for oxidation of N-free substances.

4) Heat production during oxidation of N-free substances (HEN-free): The energetic value of oxygen is calculated from the values in Table 8. It will then appear that:

a) During the oxidation of starch 0.829 l O_2 is used and 17.58 kJ is produced, corresponding to 21.206 kJ/l O_2 and RQ = 1.00.

b) During the oxidation of fat 2.013 l O_2 is used and 39.76 kJ is produced, equivalent to 19.752 kJ/l O_2 and RQ = 0.711.

The energetic value of the oxygen during oxidation of N-free substances can be calculated by means of a) and b) as follows:

	kJ/litre O2	RQ
From starch	21.206	1.000
From fat	19.752	0.711
Difference	1.454	0.289

This means that a unit increase of RQ above 0.711 will increase the energetic value of O2 by 1.454/ 0.289/1000=0.00503 kJ above the value of 19.752.Hence, the heat production during oxidation of N-free substances can be calculated as follows:

HEN-free,kJ = litre O2,N-free x (19.752 + 0.00503 x (RQN-free - 0.711) x 1000).

5) Total heat production during the oxidation of nutrients HE(RQ): The total heat production is calculated as the sum of heat production from the oxidation of protein and a mixture of fat and carbohydrate (N-free substances).

HE(RQ) = HEprot + HEN-free

6) *Correction for methane production:* For the above calculations it is assumed that N-free substances are oxidized to CO2 and H2O. This condition is not true in ruminants, in which an end-product of carbohydrate fermentation is methane. Therefore, correction must be made for this process. If the methane produced had been oxidized to CO2 and

H_2O, two litre O_2 would have been used, and one litre CO_2 would have been produced per litre of methane, since:

$$CH_4 + 2O_2 \quad CO_2 + 2H_2O$$

Therefore, the following corrections should be made: The value of O_2,N-free should be increased by 2 litres/l CH_4, and the value of CO_2, N-free should be increased by 1 litre/l CH_4. Then we get a corrected respiratory exchange and RQ values as if methane had been oxidized to carbon dioxide and water. The difference between the calculated and the true heat production is equivalent to the thermal equivalent of the methane produced (39.6 kJ/litre). Then the true heat production of N-free substances is found by deducting the thermal equivalent of the methane from the calculated heat production. Hence, the total heat production is:

$$HE(RQ) = HEprot + HEN\text{-}free - HEmethane$$

Brouwer's Equation

The above calculations were combined by Brouwer in the following equation of heat production:

$$HE,kJ = 16.18 \times O_2,litre + 5.02 \times CO_2,litre - 5.99 \times N \text{ in urine},g - 2.17 \times CH_4,litre.$$

In this function, the heat production from N-free substances is calculated from O_2 consumption and CO_2 production by multiplying the international empirical constants:

$$16.18 \times O_2 + 5.02 \times CO_2.$$

Correction for oxidized protein can be calculated as follows with approximation:

The heat production from standard protein, which is a mixture of all amino acid is 18.4 kJ/g, corresponding to 115 kJ/g N. Using the equation, for N-free substances we get: 16.18 x 0.957 + 5.02 x 0.774 = 19.37, corresponding 121 kJ/g N, which is 5.99 kJ more than the true heat production.

Correction for methane production in ruminants can be estimated from the fact that oxidation of one litre methane involves an oxygen consumption of two litres and the production of litre of carbon dioxide. When this correction is used in the equation for heat production of N-free substances, we get a lower heat production owing to the incomplete oxidation of methane: 16.18 x 2 litres O_2 + 5.02 x 1 litre CO_2 = 37.38 kJ/litre CH_4. However, the production of 1 litre CH_4 causes an energy loss of 39.55 kJ. Therefore, a difference of 2.17. kJ (39.55 - 37.38 = 2.17) per litre CH_4 has to be inserted as a correcting factor.

Other Methods

The methods for estimation of heat production and retention of energy described above require the use of respiration chamber. But attempts have been made to measure energy retention in a simpler way, based on the fact that the weight gain of an animal is the result of retention of protein, fat and minerals.

The day-to-day retention of protein and ash can be determined by balance trials. If the weight of water in the living animal can be measured, the weight of stored fat can be estimated by subtracting the lean body mass. In practice, total body water can be estimated by so-called 'dilution' techniques, in which a known quantity of a tracer substance is injected into the animal, allowed to equilibrate with the body water, and its equilibrium concentration determined. To provide valid data, a tracer must equilibrate rapidly and completely in all body water compartments, and must be capable of being measured accurately in body fluids. Several substances including antipirine, urea and hydrogen isotopes have been used to measure body water. Hydrogen isotopes have been the most throughly researched body water traces. Both deuterium oxide (D2O) and tritiated water (TOH) cross barriers at the same rate as water and distribute evenly in body fluids making them ideal tracers. However, there are some draw-backs, firstly some overestimation of body water and secondly the difficulties in quantitation of ingested water and that contributed by conception products.

Direct determination of muscle or body protein using body potassium measurement, creatinine excretion, neutron activation analyses, and 51Cr-labelled red blood cells has also been evaluated in body composition studies. The most extensive research has been attributed to potassium either by using whole body counting (40K) or isotope dilution (42K). Both techniques are based on the presence of potassium as the major intracellular cation, however, they are still expensive and not enough precise. Measurements of creatinine excretion should, in theory, be proportional to muscle mass, but there is a large variation in creatinine excretion both between and within days, resulting in inconsistent results.

Computer tomography and nuclear magnetic resonance (NMR) have been used to provide cross-sectional images of rabbits, chickens and pigs. Although, until now both methods are very expensive they have several application in studies of body composition in live animals.

In growing animals, to obtain better insight in the composition of the live weight gain, the comparative slaughter method is sometimes used. The slaughter method involves the slaughter and individual chemical analysis of a control group at the start of the experiment. Another group is fed a weighed and analysed diet for a given period and are then slaughtered and analyzed. The difference in their body composition from that of the control animals killed at the start reveals the composition of the gain and the total energy balance, consequently the difference between the total intake of metabolizable energy and the total energy balance corresponds to the total heat production. Accurate results are only obtained when the number of animals is large enough so the inevitable errors caused by between animal variation have not a great effect. This method is rather expensive and labours when used for cattle and pigs, but is often used for chickens. Recently near infrared reflectance (NIR) has been introduce as a method to save time and cost of postmortem body chemical analyses.

Estimation of Heat Production by Doubly-Labelled Water Method

The doubly-labelled water (DLW) method uses the principles of indirect calorimetry to measure total heat production (energy expenditure) from the turnover rates of two stable isotopes: deuterium and oxygen 18.

Labelling total body water also provides estimates of body composition and measurements of water outflow rates.

The principle of the method was determined in the 1950s by Nathan Lifson (1911- 89). It was over 30 years between prof. Nathan Lifson's (University of Minnesota) initial idea (1940-50) and dr. Dale Schoeller's realization in the early 1980's that improvements in the precision of isotope ratio mass spectrometers had reduced the cost of applying the technique to humans and larger animals.

Lavoisier stated the principles of thermodynamics and formulated the concept of energy balance. A steady state of body weight and body composition is achieved when energy intake equals energy expenditure. Therefore the energy requirements can be estimated from total energy expenditure, e.g. from heat production.

As indicated by "Kleiber's Law", humans have relatively low energy expenditure per unit body mass compared to small animals. Together with their profligate use of water, and hence high water turnovers, this tends to increase the potential errors in the method.

It is the first genuinely non-invasive method for measuring energy expenditure in free-living species, providing estimates of habitual expenditure over a time period of 10-20 days.

In principle the original experimental technique for humans was very simple: a subject merely drank a dose of DLW, and 2 urine samples were collected about 2 weeks apart. However, the DLW method has suffered from excessive optimism. The fundamental observation making the method possible is that the oxygen of expired CO2 is in isotopic equilibrium with the oxygen of body water (Ritz and Coward, 1995). A human or animal loaded with 18O-enriched water displays a labelling of expired CO2. This reaction is catalysed by an enzyme carbonic anhydrase . Hence, in a subject loaded with isotopically labeled water, the disappearance of 2H from body reflects water output, whereas the disappearance of 18O represents water and CO2 outputs. The technique involves enriching the body water of a subject with an isotope of hydrogen and an isotope of oxygen, and then determining the washout kinetics of both isotopes as their concentrations decline exponentially toward natural abundance levels.

The concentration of the 2H isotope, nearly all of which remains associated with water molecules, decreases as a result of dilution of body water by new, unlabelled water (consumed) as food and drink and produced during oxidation, coupled with the simultaneous loss of labelled water via evaporation and excretion. The rate constant for *H is derived as the slope of ln enrichment against time (Fig. 9) and is a measure of the rate of water movement through the subject. Most of the 18O in a labelled subject is lost as water, but some is also lost as CO2 because CO2 in body fluids is in isotopic equilibrium with body water due to the action of carbonic anhydrase present in red blood cells and elsewhere. Thus the slope of the washout line representing 18O is steeper than the line for 2H (Fig.9), and the difference between slopes represents CO2 production. This indirect measure of metabolic rate may be converted to heat production by incorporating knowledge, or estimates, of the substrates being oxidized.

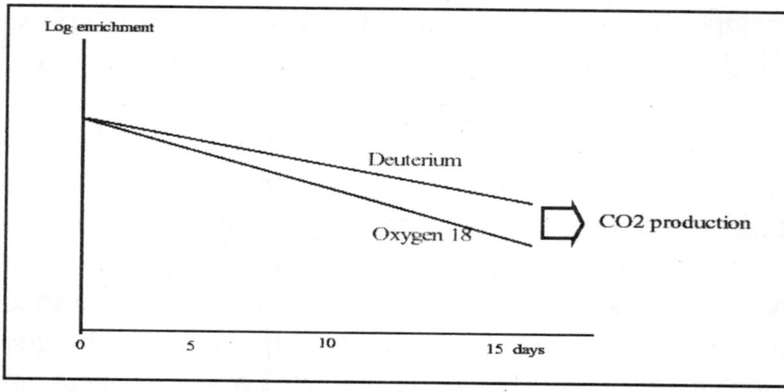

Fig. 9. Isotope disappearance curves, log transformed data

Partition of Energy in the Body

When indicating the energy metabolism of animals, we distinguish between the following physiologically determined concepts:

Gross energy (GE) is the amount of energy manifested as heat when the feed is completely oxidized. GE content of feed is measured in an apparatus known as a bomb calorimeter, which consists of a combustion chamber (the calorimetric bomb) resting in an insulated tank of water. A dried, weighed feed sample is placed in a platinum cup in the bomb. The bomb is closed hermetically, and 25-30 atmospheres of oxygen are admitted under pressure, after which the nickel wire is ignited electrically, and the feed sample burns. Owing to the high oxygen pressure, the combustion process is explosive and complete. The quantity of heat produced, calculated from the rise of water temperature, corresponds to the gross energy of the feed. The content of energy in faeces, urine and in animal products like milk or eggs is determined in the bomb calorimeter as well.

Gross energy of different materials varies, but typical values are for protein 23.86, carbohydrate 17.56 and fat 39.76 kJ/g. The differences between these nutrients primarily reflect the C: H ration and the O and N contents. For example glucose, $C_6H_{12}O_6$, has 1 atom of oxygen per atom of carbon whereas a fat molecule, for example glycerol trioleate, $C_{57}H_{104}O_6$, has 6 atoms of oxygen per 57 atoms of carbon. Thus fat requires more oxygen for oxidation and gives off more heat in the process.

Digestibe energy (DE) is the proportion of the gross energy of a feed which is apparently digested. DE = GE - FE; gross energy minus energy in faeces (FE) measured in a bomb calorimeter. Digestibility trials to determine the digestible energy of a feed or a feed ration are performed according to the same principles as determination of the digestibility quotient of the nutrients.

Metabolizable energy (ME) is the amount of energy available to the animal for body functions (maintenance, growth, production of milk and eggs, etc.). Metabolizable energy represents the energy actually capable of transformation to other forms of energy in the body, regardless of type of transformation, and whether the transformations are of any use to the animal. ME = DE - UE - CH_4E; digestible energy less urinary energy (UE) measured in a bomb calorimeter, and less methane energy (CH_4E), measured by respiration trials. ME can be measured according to the same principle as digestible energy, but digestibility trials have to be supplemented with quantitative collection of urine. Determination of ME in ruminants involves measurement of the methane produced by the decomposition of carbohydrates in the forestomachs. It may be assumed that a full-grown cow on a reasonably large feed ration produces 500 litres of CH_4/day, which is equivalent to 500 l x 39.6 kJ = 19.8 MJ, or 10% of the gross energy. In monogastric animals, however, the methane production is much less; in pigs, for examples, the methane energy is only 1-2% of the gross energy.

Heat energy (HE) is the energy produced by the total metabolism in the body. It is a sum of energy derived from oxidation of carbohydrate, fat and protein, and fermentation processes. Total heat energy includes heat produced by maintenance processes (HEm) and the so-called heat increment (HI).

Heat increment (HI) is the heat produced by the metabolism of feed and the heat produced by the conversion of the metabolized nutrients into animal products, e.g. protein and fat retained in the body, eggs, milk etc.

The magnitude of the heat increment depends on the amount of feed ingested and the composition of the diet. HI increases with increasing energy supply (amount of feed), because the transport and digestion of feed in the alimentary tract requires energy. For example, HI constitutes about 20% of the supply of 30 MJ metabolizable energy to cattle, whereas when the ME supply is 90 MJ, the HI is about 30%. The heat increment also varies from feedstuff to feedstuff, depending on structural differences, mainly owing to varying fibre contents. For concentrate feeds HI ranges from 15-25% of ME, whereas the range is 35-60% for roughages.

All body functions result in a certain heat production, corresponding to the amount of energy used in the anabolic processes. This heat production forms part of the heat increment and depends on the extent and type of the anabolic processes. In case of fat retention, for example, the ME is utilized up to 75%, i.e. 75% of ME available for fat retention is transferred to the energy retention in fat, while the remaining part is included in the total HI and is lost as part of the heat production. Similarly, the utilization of energy in protein retention is about 50%, that is, 50% is lost as heat to the surroundings.

Retained energy (RE) is the amount of energy retained in the body and/or in milk and eggs (so called energy balance). RE = ME - HE; metabolizable energy less heat production. RE in the body or in eggs is mainly energy retained in protein and fat, while in milk a substantial amount of energy is also retained in carbohydrate.

Net energy (NE) is the energy of a feed which is available to the animal for different life processes; NE = GE - (FE + UE + CH4E + HI). The deduction of the heat increment of a feed from its metabolizable energy gives the net value of a feed.

The net energy represents the quantity of energy which is equivalent to the body function caused by the feed concerned, i.e. one kJ net energy of a feed is equivalent to one kJ net energy of a body function, regardless of whether the net energy of the body function is found in a retained substance or is released as heat caused by the maintenance processes. The net energy value of the same feedstuff may differ according to the species of animal by which it is consumed and the purpose for which it is used. The total NE can be partitioned between net energy of retained substances, e.g. the sum of energy stored in protein, fat and carbohydrate, and NE associated with maintenance processes, eg. the amount of heat produced by maintenance processes.

The partition of gross energy from the feed into digestible, metabolizable and net energy is summarized in Fig.10.

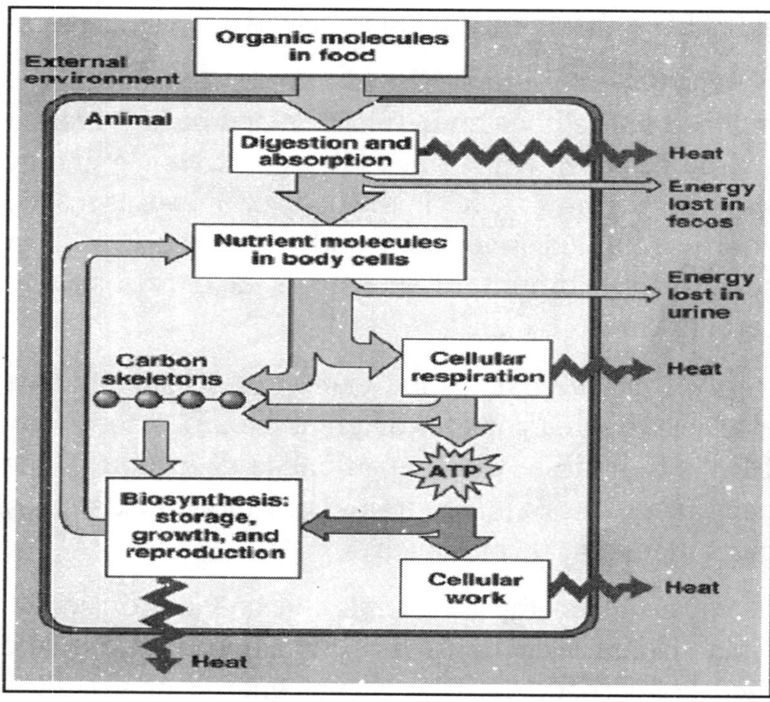

Fig. 10. Bioenergetics of an animal: an overview.

The following diagrams illustrate the partition of gross energy from feed in the animal body. The presented values are examples which are attributed to a substantial

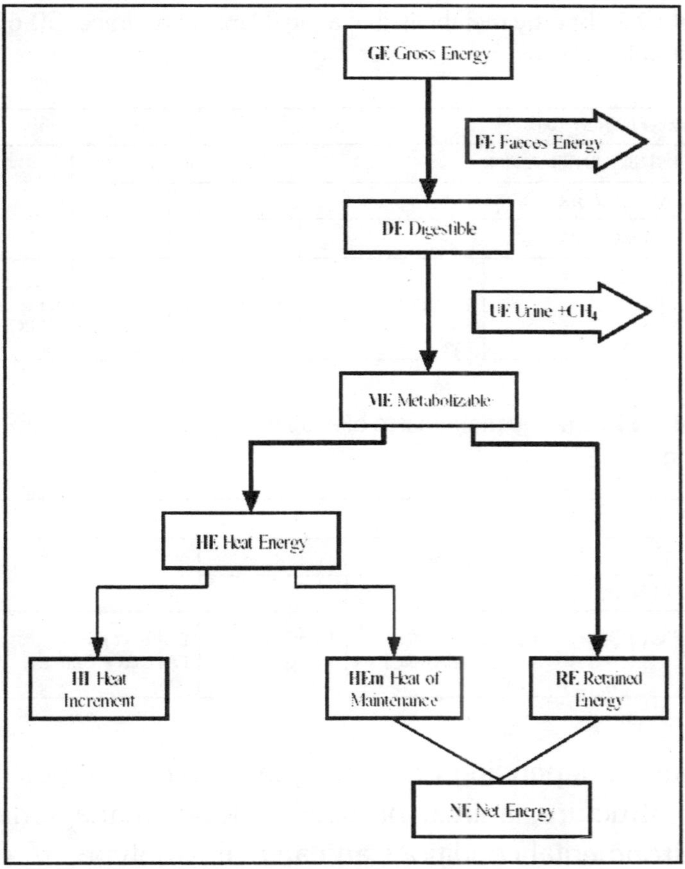

Fig. 11. Energy partition in the body.

Table 10. Energy metabolism of a dairy cow (RDM, 25 kg, 4% milk).

<table>
<tr><td colspan="6">GROSS ENERGY 100%</td></tr>
<tr><td colspan="5">DIGESTIBLE ENERGY 75%</td><td rowspan="3"></td></tr>
<tr><td colspan="3">METABOLIZABLE ENERGY 60%</td><td colspan="2" rowspan="2"></td></tr>
<tr><td colspan="2">NET ENERGY 60%</td><td></td></tr>
<tr><td>ENERGY FOR MILK PROD.
25%</td><td>MAIN-TENANCE
15%</td><td>HEAT INCREMENT
20%</td><td>METHANE
10%</td><td>URINE
5%</td><td>ENERGY OF FAECES
25%</td></tr>
</table>

Table 11. Energy metabolism of a pig (Danish Landrace, 50-60 kg body weight).

<table>
<tr><td colspan="5">GROSS ENERGY 100%</td></tr>
<tr><td colspan="4">DIGESTIBLE ENERGY 80%</td><td rowspan="3"></td></tr>
<tr><td colspan="3">METABOLIZABLE ENERGY75%</td><td rowspan="2"></td></tr>
<tr><td colspan="2">NET ENERGY 60%</td><td></td></tr>
<tr><td>ENERGY FOR GROWTH
35%</td><td>MAIN-TENANCE
15%</td><td>HEAT INCREMENT
20%</td><td>URINE
5%</td><td>ENERGY OF FAECES
20%</td></tr>
</table>

Table 12. Energy metabolism of a laying hen (Shaver St., 80% egg laying).

<table>
<tr><td colspan="4">GROSS ENERGY 100%</td></tr>
<tr><td colspan="3">METABOLIZABLE ENERGY70%</td><td rowspan="2"></td></tr>
<tr><td colspan="2">NET ENERGY 50%</td><td></td></tr>
<tr><td>ENERGY FOR EGG + GROWTH
25%</td><td>ENERGY FOR MAINTENANCE
25%</td><td>HEAT INCREMENT
20%</td><td>ENERGY OF FAECES + URINE
30%</td></tr>
</table>

variation depending on feeding level, feed composition and structure, animal production level, management, environmental conditions and animal genotype.

—★★★—

Chapter 7

Specific Dynamic Action and Efficiency of Productive Processes

Specific dynamic action, basal metabolism and exogenous nitrogen

The physiologic fuel value, or the metabolizable energy, of a diet is composed of (1) net energy, or physiologically free energy, comparable to ▲ F in thermodynamics and (2) specific dynamic action abbreviated to SDA, which is a tax to entropy, comparable to ▲ S in thermodynamics, an extra heat increment incident to the nutritive process in total, including the energy cost of excretion of the end products.

One may say that the resting heat production of an animal in a thermoneutral environment is made up to two components: (1) basal metabolism, which is the irreducible energy cost of maintenance during complete rest, and (2) SDA, an energy waste incident to food utilization. Basal metabolism is, therefore, the resting energy metabolism in a thermoneutral environment in post-absorptive condition, uncomplicated by heat increments incident to food utilization or to low or high environmental temperatures.

Basal metabolism is a convenient starting point after measuring the various heat increments, a s heat increments of : fever, feeding, lactating gestating, working, keeping warm in cold weather, and so on.

Endogenous nitrogen excretion is similarly a convenient staring point for measuring various nitrogen – excretion increments as, for example, those due to food ingestion, and it is defined, empirically, as the lowest level of nitrogen excretion attained after an empirically defined time interval on a low –nitrogen but otherwise complete diet.

Both the basal energy and endogenous nitrogen levels are useful base values for computing the energy and nitrogen costs of maintenance. One method of measuring the SDA of protein or of amino acids in by the ratio of the heat increment above the basal level to the nitrogen excretion above the endogenous nitrogen level following as single meal. There is nothing particularly edifying about the term : Specific dynamic action outside of its historical connotation,. Instead of the SDA designation, others are used, including calorigenic effect of food, heat increment of feeding, thermogenic effect and thermal energy. According to Rubner's definition of SDA: if the postabsorptive energy expense of maintenance of a mature normal animal (dog) at thermal neutrality is 100 Cal a day , then if the animal consumes 100 Cal in the form of meat, its heat production will increase to 131 Cal for the day. The extra 31 Cal is the SDA of the 100 Cal of meat. SDA of lean meat is 40 % of the intake energy: of fat, about 15 percent of sucrose, about 6 percent. that is , for example, of 115 Cal of physiologic fuel value (metabolizable energy) ingested in the form of fat, about 15 Cal is the entropy tax and 100 Cal is the physiologically free energy employed for maintenance or related useful purposes.

The contradictions in the literature on the SDA values of a given nutrient are due to the fact that the magnitude of the SDA of a nutrient is not a constant but on the contrary, is highly variable depending on many circumstances, such as on whether or not the nutrient is stored or otherwise usefully employed and this in turn depends on the balance between nutrients, plane of nutrition, environmental temperature, muscular or other productive activity, age and so on.

Moreover, every amino acid has a characteristic SDA, a characteristic speed of utilization and catabolism, which influence the real or apparent end result. Thus, the SDA of the oxidative deamination of alanine is much greater than of the hydrolytic deamination of arginine. The liberation of nitrogen from the imidoazol ring of histidine is no oxidation. The SDA also depends on the deaminized residue.

The SDA value, moreover, depends on the reference base and nuts employed, whether it is represented in terms of heat produced to total nitrogen excreted, to nitrogen excreted above the endogenous, or above the maintenance level; to grams of nitrogen, or to equivalents of nitrogen, and so on. By way of illustration, let us discuss the fate of the amino acid alanine when it is not stored in the body either because of excess ingestion or because of unbalance of the amino acid. In such case the amino acid is deaminized, producing urea, which is excreted by the kidney, and sugar, either stored in the body, converted to fat, or oxidized. The deamination may occur by three different methods(1) oxidative; (2) simple, (3) hydrolytic.

(1) $CH_3 . CH (NH_2).COOH + ½ O_2 = NH_3.CO. COOH$? urea + sugar

(alanine) (pyruvic acid)

(2) $CH_3.CH(NH_2).COOH = NH_2 + CH_3.CO.CHO$?Urea + sugar

(pyruvic acid)

(3) $CH_3.CH(NH_2).COOH + H_2O = NH_3 + CH_3.CHOH.COOH$? urea + sugar

(lactic acid)

In the first of the three reactions oxygen is consumed and , consequently heat is produced, equivalent to 4 Cal per g N catabolized. In the formation of urea form NH3, there is also heat production, about 4 Cal pr g N involved, responsible for perhaps 20 per cent of the SDA of protein.

Then, too the urea and other end products are excreted by the kidneys at a considerable energy cost, perhaps 1 to 2 Cal per g N excreted, accounting perhaps for 25 percent of the SDA of the catabolized protein , attributed to kidney work. The efficiency of the kidney is estimated to be 2 to 5 percent, 98 to 95 percent of its energy expense being wasted as heat only, which heat is part of the SDA heat.

Another methodologic factor contributing to the apparent variation in SDA of proteins due to the failure to continue metabolism measurements, of both energy and nitrogen, until after the nitrogen and heat production return to the staring base level . there are great differences in the rates of catabolism of different amino acids, different diets , in different individuals, under different conditions, and it is no easy matter to obtain physiologically equivalent values of the nitrogen and energy involved in given SDA processes. Some transformations of amino acids, as of glutamic acid to urea (and other waste products) and to glucose are exothermic; others are endothermic, therefore, not realizable spontaneously, and requiring the driving force of a coupled exothermic reaction ,and the SDA represents in the difference between the energy produced and the used in the endothermic reaction. There is considerable disagreement among the experts on all matters relating to SDA.

Carbohydrates undergo intermediate or preparatory states, for example, phosphorylation , on their path to storage (glycogen deposition or oxidation) which changes involve a heat loss, the SDA. So do fats. Moreover, the conversion of fat to glucose involves an energy loss of about 21 percent although the difference between the efficiency of fat and glucose utilization for work is only 11 percent. Furthermore, physical changes, such as solution, osmotic pressure and so on , involve additional heat liberation. The SDA of carbohydrates and fats may thus likewise be explained thermodynamically.

Plane of nutrition and SDA

It is obvious that the SDA per unit food, especially of protein, will vary with the amount of nutrient taken and with the condition of the animal. Thus milk in moderate amounts ingested by a suckling animal will have an inappreciable SDA because the nutrients, especially the proteins , are largely stored, and the stored nutrients as such have no SDA. Likewise, a breakfast small enough so that. The SDA of the dietary nutrients merely cancels that of the body nutrients which would be mobilized if the small meal were not taken, experts no appreciable extra SDA .

From above statement, it is evident that when maintenance is used as reference base, the carbon and hydrogen part of the amino acid is the major controlling factor in the level of SDA.

Nutrient unbalance and SDA

The absence of a limiting component from the metabolizable nutrients results in dietary unbalance and tends to increase the SDA. Such an unbalance may be illustrated by the absence of one of the ten odd limiting amino acids from the dietary protein . As a result of the unbalance, the animal will be unable to synthesize its characteristic tissues or other products from the remaining amino acids, As the other amino acids are not utilized for protein synthesis, they are deaminized with consequently associated SDA as previously explained.

But other limiting nutritional factors may have a similar effect. Thus the lack of common salt reduces the utilization by chicken of the metabolizable energy of corn ration. Several reports indicate similar declines in utilization of metabolizable energy when phosphate is the limiting factor. While the influence of such deficiencies on SDA remains to be worked out it is reasonable to assume that the SDA is increased because, by definition, the metabolizable energy is composed of net energy (the part utilized for maintenance and production) and SDA (the part that is wasted), and decreased metabolizable energy utilization necessarily means increased SDA.

Plane of Nutrition, the Principle of Diminishing Increments and Efficiency

The principle of diminishing increments

If the dietary metabolizable energy is perfectly balanced with regard to all nutritional needs and if it is fed below the maintenance level, it merely replaces what the liver would have delivered to the blood for maintenance, and the SDA (the extra heat increment incident to food utilization process in total) will be inappreciable; the SDA of the diet merely replaces that of the body nutrients, which would otherwise have been mobilized for maintenance needs. Above the maintenance level, however, the SDA progressively increases, or the net energy (the total food energy, less SDA and other food- utilization energy expenses) decreases for successive increments of food intake. The tendency for decline in utilization of successive dietary increments is an example of a ubiquitous phenomenon often generalized by the principle of diminishing returns, or diminishing increments, or the Wber-Fechner law, or the law of mass action .

This principle has been long recognized: "increase in yield can not be proportional to the fertilizer supplied, but it rises at a far smaller rate". "the man blessed with plenty of this worlds goods requires a correspondingly larger increase in his good fortune than does the poor man in order to derive the same amount of pleasure". "The more additional means are expended, the less striking is the effect, which stretches out to almost vanishing fineness".

The principle of diminishing increments and the net energy values of cattle and rabbit feeds

The net energy per unit food is the gross energy less the energy losses associated with its utilization. From the law of the minimum, food value table notwithstanding, the net energy per unit food is not a constant but varies with many circumstances, such as the balance between nutrients in relating to their use in the body, amount of roughage, environmental temperature, endocrine constitution of the animal, and plane of nutrition . Brody and Procter have investigated mathematically, from the viewpoint of the principle of the diminishing increments, the influence of the plane of nutrition, or amount of feed ingestion, on the digestible, metabolizable and net energy values of steer and rabbit rations.

The greatest increase in energy loss with increasing plane of nutrition appears to be due to the SDA. It rises from about 3 percent on the 0.5 maintenance ration to about 20 percent on the 3 maintenance ration. The absolute loss is greatest for the fecal energy, but the range in loss is only form 26 to 32 percent. Even fasting animals pass considerable fecal material, since undigested feed constitute but a part of the feces. The methane percentage loss in steers is relatively constant, about 9 percent of the gross energy. The urinary percentage loss of course, depends on the nitrogen intake. In this case it raged form 3 to 5 percent.

The net energy percentage (of the gross-energy intake) declines for Forbes' steers from abut 54 percent at 0.5 maintenance to about 40 percent at full feed, and for Mitchell's from about 65 to about 42 percent . the declines in digestibility and metabolizability with increasing plane of nutrition are less conspicuous.

The principle of diminishing increments and food consumption during growth

As the animal grows larger, its maintenance cost in comparison to weight gain increases and therefore, the energetic efficiency of growth decreases that is , as the animal approaches mature weight, the successive increments in body weight decrease per unit food intake ; finally , growth virtually ceases while food consumption continues for maintenance alone.

The principle of diminishing increments and milk production

There is two point of diminishing increments in milk production (1) that of energetic efficiency with increasing milk production level associated with inherent differences in lactation capacity on a standard ration , and (2) milk yield with increasing feed consumption in the same animals in which feed consumption is deliberately controlled by adding more or less of grain to the ration. The gross energetic efficiency, which carries the maintenance burden, never be as great as the net efficiency, which is not thus burned with the maintenance tax. With increasing milk yield the gross efficiency approaches the net efficiency as limit. From its asymptotic nature, this approach occurs at ever decreasing increments.

The causal mechanism actuating the decline in milk increments with successive units of grain intake in the same animal is of a more complex nature. There is first the decline in feed utilization, in net energy per unit feed with increasing feed intake. Secondly, there may be a decline in the energetic efficiency of the entire milk – producing mechanism with increasing lactation rate above a certain production level similar, for example to the decline in the energetic efficiency of an auto mobile with increasing driving speed above about 30 miles an hour . The gross energetic efficiency of milk production tends to decline if increased milk production is brought about by feeding beyond certain levels.

The gross energetic efficiency of milk production may be estimated by dividing the caloric equivalents of the milk yield (pound FCM x 340) by the caloric equivalent of the feed consumed (pounds TDN x1814). The net energetic efficiency of milk production may be similarly estimated by dividing the calorie equivalent of the milk yield by the calorie equivalent of the feed consumed above the maintenance level. There is of course, a large error in estimating maintenance –distinct from production – of a producing animal, and for this reason the net energetic efficiency of milk secretion is less close to reality than is the gross efficiency.

The dairy animals produce about 80 percent as much milk energy on an exclusive, good, roughage ration as on roughage ration supplemented with about 1lb grain per 6 lb milk (FCM), and that is generally pays to feed at least 1 lb gain to 5-6 lbs milk.

Why is an *ad libitum* roughage ration inferior to a roughage and grain ration? Probably because of physical limitations in handling sufficient roughage to supply the needed net nutrients for a high level of milk production , since one pound of grain is equivalent in net energy to several pounds of roughage. Hence the greater the lactational capacity of the cow the greater should be the ratio of grain to roughage. It is moreover, probable that a pound of TDN in the form of grain has a higher nutritive value than it has in the form of roughage. In other works, roughages may be over valued when represented in terms of the conventional TDN. The general conclusion follows that the greater the lactational capacity of the cow the greater the effect of the grain supplement on milk production. This is an illustration of a broad principle: good environment is good only in relation to the organism's ability to benefit thereby.

The principle of diminishing increments and egg production efficiency

The gross efficiency of egg production increased at diminishing increments with increasing egg production level associated with inherent differences in production capacity. The mechanism of this relation is similar to that of increasing gross efficiency of milk production at diminishing increments associated with inherent differences in lactational capacity, explained in the preceding section.

While no data are available, there is no doubt that egg production is related exponentially to feed consumption in the same manner that milk production is associated with feed consumption, that is, in accord with the principle of diminishing increments.

The principle of diminishing increments and muscular work efficiency

The gross efficiency of muscular work increases with increasing work rate, but at diminishing increments, in the same manner as the gross efficiency of milk or egg production increases with increasing productive level and for the same reason : the greater the speed of work the less the maintenance cost per unit product, approaching net efficiency (not including maintenance) as limit.

The fact that the gross efficiency of muscular work increases with increasing rate of work does not mean that it is desirable to work at a high rate; increasing the rate of work beyond a reasonable level tends to injure the animal and shorten its productive life. There is, above a certain productive level, an incompatibility between immediate and long-range efficiency. The same appears to be true of other productive processes such as milk and egg production; the span of the productive life is an important element in the overall production and profit pictures.

—★★★—

Chapter 8

Energetic Efficiency of Growth and Work Processes

While there is no sharp dividing line between the two, it is convenient to divide agriculturally productive processes in to two general classes (1) work and (2) growth, including growth of fat, milk, egg and wool, etc.

Energetic efficiency of muscular work and the maintenance complication

The theoretically maximal efficiency of the Carnot heat engine is, under the usual temperature conditions, about 50 per cent the maximal efficiency of actual heat engines is up to percent for steam engines, up to 25 per cent for gas engines, and 40 per cent for diesel engines. The average efficiencies are, of course, below maximal estimates.

Animals, however, operate not by virtue of changes or differences in temperature and pressure, as do heat engines, but by chemical energy operating at constant temperature and pressure and the efficiency of isothermal converters, as of some electric transformers, approaches 100 per cent under certain conditions. Thus the efficiency of a slow charging and discharging cadmium standard cell is almost 100 percent. Furthermore, the free energy, ?F, of the major body fuel, glucose, is practically identical with its heat of oxidation, ?H, at constant temperature and pressure. The efficiency of muscular work might, therefore, be expected to be much higher than that of a heat engine.

Actually, however, the maximal efficiency of muscular work in our horses, after deducting the ' wastes" for maintenance and walking without load, is observed to be o higher than the maximal efficiency of the Cannot engine, not over 40 per cent.

The energy is expended by the animal includes the following items:

1. Work accomplished.
2. Basal metabolism
3. Heat increment of feeding (SDA)
4. Standing, above that of lying
5. Walking without load, above that of standing
6. Overcoming internal resistance (viscosity of muscle colloids)
7. Overcoming external resistance (wind, contact of feet with ground, etc.)

8. Useless incidental movements associated with pulling
9. Additional internal work of circulation, respiration, excretion, muscle tension, etc.
10. Recovery processes.

To simplify matters, the oxygen consumptions measured during (A) standing, which covers items 2, 3 and 4 (b) working without load, item 5 and (C) pulling the load, tem 1. Item 10 is insignificant under the given conditions.

Energetic efficiency of growth and the maintenance complication

The growth process category of biologic transformations is largely in the nature of storage: in higher animals and man the proteins, carbohydrates, fats, minerals and vitamins in the diet are transformed into body, milk, egg, and so on, containing the same or derived substances. The situation is more complex in the case of bacteria and plant growth. The purely storage aspect of growth may be of no particular thermodynamic interest, but there is probably a work aspect to growth which is of thermodynamic interest.

Thus, the growth process involves creation and maintenance of: internal surfaces, differences in electromotive potentials and differences in osmotic pressure, especially such as are involved in secretion, excretion, absorption , and in general transformation of amorphous nutrients to complex organisms. The configuration of the living organism, as for example, the newly hatched chick, is on a quite different physicochemical level from that of the yolk, white, and shell form which the chick developed and it is, therefore, reasonable to assume that growth includes a work component involved in changing the configurations from that of food to that organism.

In addition to the "work- of growth" component, there are many irreversible free- energy losses which may be lumped together under the heading of "maintenance cost". Life is, so to speak, a flame, and simultaneously with the growth, building – up , process, there is a continuous oxidation, leveling , or catabolic process as predicted by the second law of thermodynamics. This " maintenance cost" is , perhaps , in part at least, the price paid for maintaining the thermodynamically unstable and statistically improbable complexity living, growing, evolving configuration, in the face of the opposite tendency of leveling disorganization to increase in entropy, as predicted by the second law of thermodynamics. Here speed of growth, size of body which has to be maintained, as on which speed and body size depend, and liberated catabolic enzymes enter as complicating elements in the estimation of the work of growth.

Energetic efficiency of prenatal growth

The bird, insect, or fish embryo is particularly suited for growth efficiency investigations because of the complete control of the organism, the definiteness of the nature of the nutrients and the convenient relative isolation from complicating environment. The gross energetic efficiency of growth of the chick up to hatching is obtained by estimating the combustion values of : (1)

the egg, (2) the newly hatched chick and (3) the unused part of the egg, such as the left over yolk and membranes. The gross efficiency of embryonic growth is of the order of 60 to 65 per cent and the energy loss as a work of growth and development (organizational energy), that is 35-40 per cent.

Energetic efficiency of postnatal growth

It is not possible to measure the energetic efficiency of post natal growth with the precision obtained on prenatal growth of the chick, because of the many variables influencing postnatal growth, such as differences in muscular activity, food supply and feeding habits, nature and quantity of food especially balance between nutrients, environmental temperature and so on. There are moreover, economic difficulties in making balance sheets of income and outgo of energy and especially determination of the composition of the gains.

Farmers estimate what they call efficiency – really a reciprocal function of efficiency – by the amount of feed consumed for making a given gain in weight. The terms "amount of feed" and "amount of gain" are poorly defined since the energetic equivalence of availability of feed and the energetic equivalent per unit body weight gain are highly variable.

It is instructive to quote, by way of introduction, Rubner's "law" of growth , to the effect that the amount of energy required for doubling the birth weight is the same per kilogram in all species, except man. About

4800 dietary cal are required to construct a kilogram of normal body substance containing 30 g nitrogen and 1720 cal. The energetic efficiency of the early postnatal growth is thus 1720/4800x100 = 35 percent. The following is an examples of Rubner's computations" cal in the gain, 1500; cal for maintenance, 2500; therefore, the energetic efficiency of growth is 1500/(2500+1500)= 37 percent.

Rubner evidently refers in this example to what we call net energy of the feed. Indeed, he makes no mention of the feed used only the energy in the gain and the (roughly estimated) energy cost of maintenance. The efficiency would, of curse, be much lower if the energy in the gains were divided by the energy in the food consumed and this value would vary with the fed category employed – gross, digestible, metabolizable, or net. To indicate the differences obtained by two comparable investigators, on may cite Luck, who reported that pigs retain 20 percent of their dietary calories during the first doubling of body weight, whereas Rubner stated that 40 percent of the ingested energy is retained. Lusk reported the combustion value of the gain as 866 cal per kilo, whereas Rubner stated that the combustion value of the gain as 1722 cal per kilo. Yet, both Lusk and Rubner seem to agree that the milk energy required for a pig to double its birth weight is about twice the amount of energy deposited in the body.

The gross efficiency of growth of Jersey cattle with respect to TDN consumed indicate an early efficiency near 35 percent with a decline to 5 percent as age 2 years. The efficiency of beef production within the given limits is seen to be below 10 per cent. The apparent efficiency would be reduced still further by deducting the inedible part of the body weight.

—★★★—

Chapter 9

Energetic Efficiency of Milk Production

The gross efficiency of milk production in ' good ' dairy cattle and dairy goats is of the order of 33 per cent with respect to TDN consumed (assuming 1lb TDN = 1814 cal); and this efficiency, like that of growth at a given physiologic age, is roughly independent of body weight, or even of species as such. When other conditions are equal, it is approximately the same in dairy cattle, dairy goats, and white rates. It is true that the milk production per unit body weight is greater in small than large animals, but the basal metabolism, or maintenance cost, per unit body weight is also greater in small than large animals with the net result that the two balance and the energetic efficiency is approximately independent of body weight. In this case Kleiber's rule holds. About

1/3 of the consumed TDN (lb TDN x1814) is usually recovered in the milk from good diary animals regardless of body size if other conditions are equal.

The total energetic efficiency of milk production is comparable in magnitude to the total efficiency of very early postnatal growth, but it is very much greater than of beef production at later ages. As egg is a more complex substance than milk, it requires more time for its elaboration, so that a greater proportion of the chicken's food would be expected to be used for maintenance in elaboration of egg than of an energetically equivalent quantity of milk with consequent lower total efficiency for egg than of milk production. The total energetic efficiency of egg production in good egg producers is approximately half that of milk production in good dairy animals.

It thus seems that, when other conditions are equal, the greater the structural complexity of the product the lower the energetic efficiency of its production. Milk, the least complex product, may be produced at the highest efficiency; this is followed by egg , and this by growth (meat).

Energetic efficiency of milk production

The overall economy of milk production is composed of many elements, the two outstanding of which are

1. Energetic efficiency of the lactation process.

2. Size of animals and plane of nutrition.

Gross energetic efficiency of milk production

Is the ratio of milk calories produced to feed calories consumed, including maintenance cost.

$$\text{Gross efficiency of milk production} = \frac{\text{Milk calories produced}}{\text{Feed (TDN) calories consumed including maintenance cost}}$$

Net energetic efficiency

Is the ratio of milk calories to feed calories above the maintenance cost.

$$\text{Net efficiency of milk production} = \frac{\text{Milk calories produced}}{\text{Feed (TDN) calories consumed above the maintenance cost}}$$

In addition to the overall gross and net efficiency categories as defined above, one may measure the efficiencies of the various component processes as, for example, the energetic efficiency of the mammary gland alone in assembling the milk precursors in the blood and synthesizing them into milk. The gross efficiency of the mammary gland, including its maintenance cost, is approximately 90 percent; that is, 10 per cent of the milk precursor energy entering the mammary gland was expended for the synthesis of milk from its precursors, and for the maintenance of the gland. The heat increment of mammary gland activity constitutes nearly 20 per cent of the basal metabolism. It was noted in the previous section that the heat increment of heavily lactating animals is about 100 per cent above that of the non-lactating level.

Energetic efficiency of milk production in dairy cattle

The gross efficiency of milk production of dairy cows weighing from 1000 to 1300 pounds, producing 28 to 50 lbs FCM and consuming 17 to 28 lbs TDN per day – produced milk at a gross energetic efficiency of 28 to 34 per cent; that is, 28 to 34 Cal milk were produced per 100 Cal TDN consumed. In other words, about 1/3 of TDN energy consumed by the superior cows was recorded in the milk; or the TDN energy consumed was about 3 times the FCM energy produced.

The gross energetic efficiency of milk production of "good" dairy cows is of the same order as the gross efficiency of early postnatal growth on "good" diets, namely about 30 per cent. Thus chicks and cattle grow at roughly, the same gross energetic efficiency in early postnatal life, just as large and small cows tend to produce milk with equal gross energetic efficiency.

The Net energetic efficiency of milk production with respect to TDN consumption in dairy cattle is of the order of 60 per cent. Since net efficiency does not include maintenance cost, therefore gross efficiency, which does include maintenance cost, can never reach the net efficiency of 60 per cent level; 50 % is probably as high a gross efficiency level with respect to TDN consumption as it is possible to achieve.

Energetic efficiency of milk production in dairy goats

The gross ene1rgetic efficiencies of milk production in respect to TDN ranges from 32 to 40 per cent, averaging 34.9 per cent. The conclusion of above statement is that the gross energetic efficiency of milk production in dairy goats is of the same order as that of diary cattle.

Influence of body size on gross efficiency of milk production

It is said the large animals are better milk producers than small ones because, as a rule, thy produce more milk. It is also said that small animals are better milk producers than large because (other conditions being equal) small animals produce more milk per unit body weight than large. Thus, as explained above, the average milk calories produced per kg body weights are about 175 for rats and 25 for diary cows.

However, the comparison of milk production in the preceding section shows that the energetic efficiency of milk production (the ratio of milk energy produced to food energy consumed) is virtually the same in rats, goats, humans and cattle. The indifferences of opinion concerning the influence of body size on milk production are due to differences in the reference bases employed. The fact is that when other conditions are equal, large animal produces more milk per animal than a small animal; a small animal produces more milk per unit weight than a large animal; large and small animals produce the same amounts of milk energy per unit food energy.

It is evident from the above statement of cattle, goats, rats and humans that body size as such does not influence the energetic efficiency of milk production; other conditions being equal, small and large animals convert the same percentage of good energy into milk energy.

The gross efficiency of milk production of good dairy cattle with respect to TDN energy consumed is of the order of 30 per cent. However, it appears from indirect evidence that at the biologic limit, that is, in the highest-producing champions, the gross efficiency may rise to 50 per cent. Following the psychologist's I.Q. (intelligence quotient) concept we may say that if the gross efficiency of milk production of the "average good" cows is 30 percent, and if a given champion cow's gross efficiency is 50 per cent, the L. Q. (lactation quotient) of the champion cow is 48/30 = 160. A cow having a L. Q. of 166 is a "lactation genius", just as a man having an I. Q. of 166 is an "intellectual genius".

Net efficiency of milk production

The average gross efficiency (including maintenance cost) of milk production of good dairy cattle is of the order of 30 per cent and of champion cows, at the biologic limit near 50 per cent, we next proceed to investigate the net energetic efficiency (excluding maintenance cost) of milk production. It is true that from the economic viewpoint the overall or gross efficiency is the more important; but we shall see that the net efficiency is also of considerable interest. The Net efficiency of milk production with respect to TDN consumption is of the order of 60 per cent.

The gross energetic efficiency of milk secretion, which carries the burden of the maintenance cost of the cow, can never be as great as the net efficiency, which is not thus burdened with maintenance taxes. As production increases, the maintenance tax per unit milk produce becomes less and less; therefore, the gross efficiency approaches nearer and nearer to the net efficiency level, but at decreasing increments, in accordance with the law of diminishing returns. Since the net efficiency of milk production is of the order of 60 per cent, the gross efficiency can never reach 60 per cent, but is may approach 50 per cent.

The gross energetic efficiency of milk production is apparently independent of live weight as such. It is true that the ratio of milk energy yield to body weight is greater in small animals, such as 120 lb goats, than large animals such as 1200 lb cows; but the ratio maintenance energy cost to body weight is likewise greater of the 120 lb goat than for the 1200 lb cow, with the net result that the extra maintenance cost counterbalanced the extra milk yield and the gross efficiency is unaffected by body size.

—★★★—

Chapter 10

Energetic Efficiency of Egg Production

Eggs resemble milk in most respects (evolution as food for young, composition, nutritional value, physicochemical and endocrine mechanisms), and differ in few, especially in concentration of carbohydrate, fat, calcium, iron, copper and vitamin C.

The energetic efficiency of egg production is evaluated by the same method as for milk production with the aid of the following two basic equations:

Gross energetic efficiency of egg production

$$= \frac{\text{Egg calories produced}}{\text{Feed (TDN) calories consumed including maintenance feed}}$$

Net energetic efficiency of egg production

$$= \frac{\text{Egg calories produced}}{\text{Feed(TDN) calories consumed less TDN expended for maintenance}}$$

If body weight and composition are assumed to remain unchanged, the gross efficiency with respect to TDN (total digestible nutrients) is computed from the equation (in which 1.6 and 3 are , respectively , the calories equivalents per gram of whole egg and of assumed TDN in the feed).

Gross energetic efficiency of egg production

$$= \frac{1.6 \text{ x egg grams produced}}{3 \text{ x feed grams consumed}}$$

The net efficiency (not including maintenance cost) is estimated by mathematical partitioning of the feed consumed between its three uses : egg production , maintenance and body weight gain (or loss). The partition equation for egg production is of the same form as for milk production:

$$\text{Feed} = B\ (\text{egg}) + CW^{0.73} + D\ ?\ W$$

The significance of its parameters (B, C, D) depends on the accuracy of the basic data (TDN, egg, live weight, live – weight gains), the size and homogeneity of population, range of data and method of fitting equation to data

The net efficiency, that is, the ratio of egg calories produced to TDN consumed above the maintenance and weight gain cost, is therefore given by the relation

Net energetic efficiency of egg production

$$= \frac{1.6 \times \text{grams egg produced}}{3 \times \text{gram feeds consumed for egg}}$$

The above numerical result means that under the given condition, out of 100 cal. TDN consumed above maintenance and above live weight changes as computed by us, 77caloreis were recovered in the eggs and the reminder expended for the cost of various intermediate transformations. It may be recalled that the net efficiency of milk production with respect to TDN in cows is computed to be 62 percent. The difference of between 62 and 77 per cent in net efficiencies of milk and egg production appears to be within the limits of variability of the data and equation constants. From above statement is evident that the net efficiency of egg production is only 54 percent.

The net energetic efficiency thus appears to be of the same order of egg as for milk production; that is, the energetic efficiency of transforming TDN into eggs when maintenance cost is not included is of the same order as the efficiency for milk production when maintenance cost is not included. It is therefore concluded that the ovary and oviduct produce egg form egg precursors with the same energetic efficiency as the mammary gland produces milk form milk precursors.

Influence of live weight on gross efficiency of egg production

The gross energetic efficiency of egg production rises with increasing egg production but at decreasing increments in accordance with the law of diminishing increments, approaching a gross efficiency of 15 t 35 per cent as limit, depending apparently on body weight. It appears the large fowls can not attain as high an efficiency level as small.

The egg weight and in some cases egg number tend to increase with increasing body weight up to a certain body weight level, following which egg number deceases. The body weight of maximum production is of the order of 4 lbs in Leghorn and $5^{1}/2$ lbs in the heavier breeds as R. I. reds and B. P. Rocks.

The 1 gram whole egg is equivalent to 1.6 calories and 1 gram chicken feed 3 calories TDN per gram, or 1360 calories TDN per pound.

Influence of production level on gross efficiency of egg production

The maximum efficiency of egg production should be of the order of 25 per cent. The "good" layers, good enough to have gross energetic efficiency with respect to TDN consumption up to at least 19 per cent. If a pen of bird can produce eggs at 19 per cent efficiency, it is probable that individual birds may reach 25 per cent efficiency level.

Summarizing, the energetic efficiency of egg production, like that of milk production, increases with increasing production level in accordance with the law of diminishing increments. The energetic efficiency of pens of birds at the given egg – laying competitions and ranged from about 10 to 19 per cent with respect to TDN consumed. Assuming that the net efficiency (not counting maintenance cost) of egg production is constant (about 1 lb feed per dozen eggs) these 10 per cent differences in gross efficiency are due to differences in maintenance tax imposed on the egg. The larger the number of eggs produced, the smaller the maintenance per egg and consequently the higher the gross efficiency.

Relation between the net energy category of feed and the efficiency of productive processes

The net energy of a feed is, by definition, the part transformed into the desired product: work, body tissue, milk, egg and so on. The net efficiency of a productive process, excluding maintenance cost should, therefore, be about 100 percent with respect to the net energy category of feed ingested. The net efficiency of growth should be less if the efficiency is computed with respect to the metabolizable energy category of feed (perhaps 70 per cent in cattle), and least (50 per cent in cattle) if is with respect to the gross energy category of ingested feed.

In discussing efficiency it is, therefore, imperative to specify what category of efficiency – gross or net and what category of feed- gross, digestible, metabolizable or net – is under consideration. The literature on productive processes is confusing, from the energetic-efficiency viewpoint, either because of poor specifications or because of lack of appropriate conversion factors under the given feeding conditions.

The gross efficiency (including maintenance cost) of milk production with respect to TDN consumption (assuming that 1 lb TDN is equivalent to 1814 cal) is about 33 percent.

The net efficiency of milk production, not counting maintenance, is thus seen to be nearly double that of the gross efficiency, including maintenance. It is surprising that despite great differences in gross efficiency of growth, milk production and egg production, the net efficiency, excluding maintenance appears to be roughly of the same order in all three processes. The net efficiency of gains in weight ranges from 45 to 63 percent in a cow and 52 to 84 percent in a growing pig – values close to those for the net efficiency of milk and egg production.

The differences in the gross efficiencies of growth, milk production, and egg production may reflect differences in the time intervals required to produce unit energy in the form of milk, egg and body gain, with corresponding differences in maintenance cost and consequently differences in gross efficiency time and therefore, maintenance profoundly affect the gross efficiency of all productive processes.

The gross efficiency of growth, including the maintenance complex, is 3 percent of certain autotropic bacteria, about 10 percent for yeast, 60 percent for prenatal growth, 35 percent for early postnatal growth, 5 to 10 percent for beef production in later ages, near 16 percent for egg production, and about 33 per cent for milk production. The net efficiency, which does not include the maintenance items is of course, higher by the maintenance cost, and is roughly the same for all these processes. The gross energetic efficiency of growth depends on the physiologic age interval under consideration because the older the animals the greater the maintenance taxes in comparison to the productive increment.

—★★★—

Chapter 10

References

Andre Chwalibog (1994). Physiological basis of heat production (Fire of life). The Royal Veterinary and Agricultural University, Denmark.

Brody, S. (1945). *Bioenergetics and Growth*. Hafner Publishing Company, INC. New York.

Brody, H. (1955). Organization of the cerebral cortex. III. A study of aging in the human cerebral cortex. *J. Comp. Neurol.* 102: 511-556.

Brody, S. (1945). *Bioenergetics and Growth*. Hafner Publishing Company, INC. New York.

Brody, H. (1955). Organization of the cerebral cortex. III. A study of aging in the human cerebral cortex. *J. Comp. Neurol.* 102: 511-556.

Currie W. Bruce (1992). Structure and function of domestic animals.

Crane,Frederick L. and, D. James Morré (1991). Oxidoreduction at the Plasma Membrane: Relation to Growth and Transport: Animals. Published by CRC press.

David A Harris (1994). Bioenergetics at a glance: an illustrated introduction. Published by Blackwell publishing.

Gaunt, R. (1954). Chemical control of growth in animals. In: Dynamics of growth processes. E.J. Boell (Ed.), Vol. II, pp. 85-121, New York, Academic Press.

Gruenwald, P. (1967). Growth of the human foetus. In: Advances in Reproduction Physiology. McLaren (Ed.), Vol. II, New York, Academic Press.

Hafez, E.S.E. and Dyer, I.A. (1969). *Animal Growth and Nutrition*. Lea & Febiger. Philadelphia.

Hammond, J. (1961). Growth in size and body proportions in farm animals. In: *Growth in living systems*. New York, Basic Books, Inc.

Harvey (1959). In: *Reproduction in Domestic Animals*. H.H. Cole and P.T. Cupps (Eds.), New York, Academic Press, Inc.

Joubert, D.M. (1956). An analysis of factors influencing post-natal growth and development in the muscle fibre. *J. Agric. Sci*. 47: 59-102.

Lawrence, T.L.J. and Fowler, V.R. (2002). *Growth of Farm Animals*. 2nd Ed. Sydney, Butterworth.

Needham (1933). *J. Biol. Rev.*, 8, 180.

Peter Graber and Piulio Milazza (1997). Bioenergetics. Published Birkhauser.

Rosenberg, I.N. and Bastomsky, C.H. (1965). The Thyroid. Ann. Rev. Physiol. 27: 71-106.

Schloss, E. (1911). Pathologie des wachstums, p. 9. Berlin; S. Korgen. (Quoted by Maynard, L.A., Animal Nutrition).

S. Sundara Rajan (2003). Introduction to bioenergetics. Published by Anmol Publications Pvt. Ltd.

Wallace, L.R. (1948). The growth of lambs before and after birth in relation to the level of nutrition. *J. Agric. Sci.* 38: 243-302, 367-401.

Walton, A. and Hammond, J. (1938). The maternal effect on growth and conformation in Shire horse- Shetland pony crosses. Proc. Roy. Soc. (Biol) 125-311.

Wrigglesworth John M (1998).Energy and life. Published by Trylor and Francis.

Wells, L.J. (1965). Fetal hormones and their role in organogenesis. Pp. 673-680. In: *Organogenesis*. R.L. DeHaan and H. Ursprung (Eds.), New York, Holt, Rinchart & Winston.

—★★★—